MANAGING THE CLIMATE CRISIS

MANAGING THE CLIMATE CRISIS

Assessing Our Risks, Options,
and Prospects

ROBERT O. SCHNEIDER

 PRAEGER™

An Imprint of ABC-CLIO, LLC
Santa Barbara, California • Denver, Colorado

Library of Congress Cataloging-in-Publication Data

Names: Schneider, Robert O.
Title: Managing the climate crisis : assessing our risks, options, and
    prospects / Robert O. Schneider.
Description: Santa Barbara, California : Praeger, [2016] | Includes index.
Identifiers: LCCN 2015028477| ISBN 9781440839986 (alk. paper) |
    ISBN 9781440839993 (ebook)
Subjects: LCSH: Climatic changes—Social aspects. | Climatic
    changes—Risk assessment. | Climate change mitigation.
Classification: LCC QC903 .S365 2016 | DDC 363.738/74—dc23 LC record
    available at http://lccn.loc.gov/2015028477

ISBN: 978-1-4408-3998-6
EISBN: 978-1-4408-3999-3

20 19 18 17 16   1 2 3 4 5

This book is also available on the World Wide Web as an eBook.
Visit www.abc-clio.com for details.

Praeger
An Imprint of ABC-CLIO, LLC

ABC-CLIO, LLC
130 Cremona Drive, P.O. Box 1911
Santa Barbara, California 93116-1911

This book is printed on acid-free paper ∞
Manufactured in the United States of America

Contents

CHAPTER 1

Feeling the Heat

Introduction

Americans disagree about climate change. There are reasonable things about which to disagree on the subject, but much of the American political "debate" about the science is rather pointless, given that there is not much debate or disagreement in the science itself. As one who takes seriously the science and its consensus about climate change and the set of challenges it presents for humanity, I am inclined by my interests and background to think about climate change/a.k.a global warming in the context of emergency management. At least, given my background in that area, this is the framework that leaps out to me and helps to orient my thinking on the subject. By this I mean that my interest and focus is irresistibly drawn to the implications of a warming climate for future natural disasters, our preparedness for them, our response to them, and most especially our efforts at hazard mitigation. This brings everything nicely into focus for me. It is what animates my interests, heightens my concerns, and has motivated me to write this book.

Emergency management seems for me to be a particularly appropriate context to think about global warming. Its logic speaks most clearly to me. As global warming and the challenges it represents unfold, it will, according to the reliable established scientific consensus, significantly influence the course of future natural disasters. It will change hazard risks and the vulnerability profiles for all the communities in which we live and work. In this context, it has occurred to me, for example, that emergency management must be logically expected to translate its past efforts and make them relevant for a world where a warming climate will significantly alter the intensity, frequency, and damage-related costs of future natural disasters. It strikes me also that the tilting of the climate change conversation in this direction may provide a unique context for understanding that all of us need to translate our social, economic, and political activities to make them relevant for a world where a warming climate is altering the future for humanity. That sounds like good old-fashioned common sense to me. But it is a common sense that is questioned by many, too many one might say, in the American discourse about the subject. Is the climate really warming? Are we feeling the heat? Is global warming a crisis? How should we respond? Will it really result in a disaster if we do not respond to it?

Climate scientists have concluded that enough is in fact known to document that global warming is a critical event and an important moment of decision is upon us. Many Americans are, for a variety of reasons we need to understand and will explore and discuss in this book, reluctant to accept this conclusion. In fact, some even question the validity of the science. I cannot, of course, hope to silence all of the contrarian claims made by climate change deniers in this book. Neither can I as a non-scientist comprehensively and expertly address all of the science on the subject. But as a social scientist with an interest in the scientific basis for our current understanding of global warming or climate change and its implications for sound and rational public policy, I believe that I can say a few compelling things about the intersection of science and politics. As an emergency management scholar, I believe I may have a few insights to offer regarding the societal and human impacts of a warming climate that we will, whatever our beliefs about the subject, find ourselves having to deal with as they challenge the resiliency and the sustainability of our communities. My purpose will be to share my understanding of the scientific foundation, the policy implications, and the societal and human impacts of global warming. I hope to contribute to and encourage others to engage, through what I intend to be a slightly different prism, in what I consider the most important discussion of our time. It is time to have an informed and urgent discussion about how we can intelligently manage what is clearly identified as a "climate crisis." The basic premise is that science has been able to warn us that we have a problem. It tells us with growing clarity that a crisis situation is already in motion, and should we fail to intelligently address that crisis, we may experience a genuine planetary disaster.

There are, of course, an incredible number of excellent books on the subject of climate change from a variety of perspectives, so why should I write this one, and why should anyone read it? My answer begins with the observation that too many of my students, including graduate students, indicate that they are surprised to learn that climate change is not a new or recently identified concern or development. They are genuinely surprised to learn that it has a long history in the scientific literature. Likewise, many still believe climate change to be a matter of some considerable scientific dispute. Few know that the science has been pretty much settled for quite some time now and that most of the disagreement is "political" rather than "scientific." Most do not comprehend the many decades of politics surrounding the issue of climate change and the often-disconnectedness of that politics from the science. Among those who may understand some of the science and some of the politics, I continue to find that many believe that climate change is a concern for the future but not necessarily for the present. Ultimately I find that few students understand the basic threats and vulnerabilities a warming climate presents to all life on the planet in practical and personal terms. As a result, they do not begin to really think seriously about or have any urgency regarding the need for practical responses to these threats and vulnerabilities or the public policy context necessary for adopting and implementing them.

I know that my students and their perceptions are not that much different from the public, generally, when it comes to the topic of climate change. So the intended purpose of this book is to provide an accessible and fairly comprehensive overview of climate change but with a specific focus on risks, vulnerabilities, options for response, and prospects for progress. Applying some of the general lessons available from the study of sustainable disaster mitigation and hazard resilience, *it may be possible to articulate more effectively to a broader audience the practical need to respond to climate change.* This may contribute to a reorientation of the climate change discussion to the *rational assessment of important challenges and the implementation of important practices and policy changes.* This is the discussion we need to have rather than the contentious debate about the reality of the threat that seems to define partisan political conversation. It is the discussion too few Americans know how to have. If this book contributes to such a discussion, it will have achieved its purpose. We must begin with the basic scientific reality and follow it to what is the essential but very basic truth of the matter. Climate change is happening. This is a bad thing. Humans are responsible for it. We really should do something about it. An assessment of the robust science that supports this truth makes it abundantly clear that it is time to call for necessary action in response to identifiable threats to humanity and to nature. It is time for all of us to take responsibility for climate change. It's happening right now. We must come to understand that we have a very serious problem, and it's time to respond.

Earth, We Have a Problem

Climate change is not a new invention or a passing fancy. For over one hundred and fifty years, scientists have refined their methods of study and improved our knowledge about it. Climate science is sound, but that does not mean that it is universally accepted or embraced. If you think about it honestly and reflect some on our history, science is never universally embraced when it presents us with new information that defines new problems that require substantial changes in the way we live our lives. Most of the time, the warnings of science will meet fierce resistance under these circumstances. There are always those who, with a huge financial or political stake in preserving the status quo or ignoring the problem, will promote doubt and suspicion about the science. For example, in the not-too-distant past, science produced conclusive landmark studies on the dangers of DDT, tobacco, acid rain, and ozone depletion. In each of these cases, as is presently the case with global warming, economic and political interests delayed action. Along with a more general public inclination to reject any suggestion that we must adjust our economic and material pursuits in a manner that requires us to defer profits, pleasures, or other benefits in order to manage long-term threats, they combined to reduce the felt sense of urgency to act. It took a great deal of time and long political battles had to be fought before we acted on these dangers so accurately and completely documented by sound science. Is global warming like DDT, tobacco, acid rain, and ozone depletion? Despite the human costs associated with delay, is there time to fight the political battles that must be fought before we act in response to the science? Is the climate challenge manageable in the normal course of events? Do we have time to spare before we act? Or is it a crisis already at hand requiring our immediate and urgent attention? From the emergency management perspective that influences much of my thinking and pulls me into the dialogue about climate change, that is perhaps the most relevant question.

I am old enough to remember the drama of Apollo 13. I still get a chill when I hear a replay of Commander Jim Lovell's famous and gripping message from space: "Houston, we have a problem." This message signaled a major technological disaster that led to the abandonment of the lunar mission for Apollo 13. The priorities shifted very quickly, and with very little time for debate and second guessing. The crisis at hand, with its potential to result in a predictable disaster or tragic outcome, changed everything. Survival at all costs became an immediate imperative. Work the problem. Bring them home safely. Along the way home, the Apollo cabin had to be shut down, and the Lunar Module had to be used as a backup vessel. CO_2 and oxygen problems had to be solved, course alterations had to be made, and a variety of other unexpected problems had to be resolved. There was no precedent and no manual for the crisis at hand. The Apollo 13 heroes, including everyone working on the ground team in support of the three astronauts, recognized and dealt immediately and effectively with the crisis. They did not have a perfect knowledge of what had gone wrong, of how much damage had been done, or whether what they were attempting to do would work. But they recognized the crisis and the necessity to act. The rest, as they say, is history. Good history, at that. They prevented a disaster.

It can be said that climate scientists are, on the basis of their work, sending us a gripping message as well. It too should send a chill up our spines. "Earth, we have a problem." A growing number of reliable indicators signal that a climatological disaster may be in the making. Our life systems (ecosystems) are in danger. Spaceship Earth is sustaining measurable damage, and the risks to humanity are escalating rapidly. Human survival is a real concern. There is no backup ship for the earth in this case. Some would suggest that it is time to work the problem. The mission is changing. There is no precedent, and there is no manual. Nothing in human history has adequately prepared us for what is to come. However, there is science, and there is time to resolve the crisis before a disaster or catastrophe of unthinkable proportions occurs. But even with the advantages of some very solid science, how much time is there really? Do we have the right sense of urgency about acting? Do we recognize the crisis? Are we refocusing our attention and preventing a disaster?

As the Apollo 13 crisis unfolded, the dominant message of the leadership at NASA's Mission Control was "failure is not an option." As the climate crisis continues to unfold, the message from Mission Control—government policy makers in this case—is quite different.

There is considerable reluctance to identify the problem as a crisis. In some cases, there is reluctance to accept the science. In some cases, there is an acceptance of the science, but not an agreement that there is any danger to be navigated or crisis that requires an immediate or urgent response. In still other cases, there is an assumption that there is an abundance of time and, without significantly altering or sacrificing immediate objectives or angering powerful economic interests, we can piecemeal something together at some indefinite future date. The net result is that, with respect to what we shall define as a climate crisis, the message from Mission Control is "urgent action is not an option" in the political realm at the present time. America is not alone in its slowness to respond to the warnings of science. Political inaction, like climate change, is global.

Global warming deniers are, as we all probably know, unmoved by the body of peer-reviewed science that supports the indisputable conclusion that global climate change poses a serious challenge that will test the limits of nature, impact human populations, and adversely influence markets and economies. It can be said also that most of us, including those who embrace the scientific consensus about climate change, are unmoved to act with a genuine sense of urgency until the effects of a changing climate impact us directly where we live. But all of us, the deniers included, are beginning to feel the heat. Whether we accept or reject the message of science—"Earth, we have a problem"—we have had an expanding number of climate-related experiences over the past decade that have made us all sweat.

According to the National Climate Data center, 2012 was the warmest year on record in the United States up to that time. Especially interesting to me when this report came out was the ratio of heat records to cold records. In 2012, we saw 362 all-time high temperatures across the nation and zero all-time record lows. Most of the nation (61 percent of all counties) experienced moderate to exceptional drought conditions, and there were a record number of serious wildfires as the United States experienced its driest year since 1988. (1) People were feeling the heat. They would continue to feel it. As it happened, 2013 surpassed 2012 as the warmest year on record, and in the fall of 2014, the National Oceanic and Atmospheric Administration projected that 2014 would surpass the 2013 record (it did), and 2015 is looking like it may make the 2014 record a short-lived one. Despite colder winter temperatures in some locations, the first three months of 2015 were recorded as the warmest ever January to March, globally. But that does not mean there is any broad public agreement that a crisis of any sort is at hand or a disaster is unfolding. Scientists may warn, for example, of plausible planetary-tipping points and emphasize the need to address the root causes of how humans, their technology, their development practices, and their lifestyles are forcing biological changes. (2) We will easily and without much thought ignore any such warnings unless or until we feel the impact personally. But we do take note, for at least awhile, when we are directly impacted by the heat. Such has been the case as extraordinary global heat events have multiplied over the first years of the 21st century. Even a casual review of our recent experiences should tell us that something is indeed changing.

Extraordinary Heat Events

Short-term heat waves cause spikes in the demand for electricity. Almost everyone knows this. On the afternoon of August 14, 2003, residents and workers in New York City felt the heat quite dramatically as they witnessed and experienced one of the most massive power outages in American history. (3) It was not only the heat wave that brought things to a standstill. There were also serious infrastructure issues at play. The interaction of extreme temperatures with the constructed, aging, and poorly maintained electrical infrastructure demonstrated how nature's extremes can combine with human shortcomings quite effectively to create the potential for disaster.

The heat wave experienced in August of 2003 caused a spike in the regional demand for electricity. New York City was adversely impacted in no small part because of a failure by the United States over many decades to invest in upgrading its transmission infrastructure. Many regions of the country rely on an incredibly small number of transmission corridors to carry excessive electrical loads during periods of high demand. It was in one such corridor in rural Ohio that the demand grew so incredibly intense as to cause the New York City power outage. The high-voltage power lines in this particular corridor began to physically sag as a result of the intense heat associated with extreme electrical loads. This, in turn, caused the transmission lines to come into contact with tree limbs, short circuiting and disabling them. A sudden drop in electrical transmission to nearby cities created a surge of power from the east to compensate. This caused a power void in the New York region that required a massive surge from suppliers to the west. Cascading power surges can cause damage to generating plants and, in order to avoid more dire consequences, the regional system had to be shut down entirely as a precaution. (3) New Yorkers were unexpectedly and without warn-ing brought to a standstill by a massive power blackout. As they felt the heat, the rest of the nation watched in air-conditioned comfort. The problem subsided over the next 24 hours, and the power-generating plants were gradually brought back online. The regional electrical grid was restored.

Few of us may remember or think much about the 2003 New York City blackout, but it is perhaps prudent to suggest that we should remember. Most American cities are today experiencing more than 20 days (three weeks) per year of extreme temperatures that are hot enough to cause public health concerns. The number of such days has been and is expected to continue increasing. With more intense heat in American cities comes the increased likelihood of heat-related ill-nesses and increased stress on an electrical infrastructure in desperate need of repair and modernization.

Europeans felt the heat in the summer of 2003 as well. They expe-rienced temperatures so extreme for them that it shattered all records kept in over 300 years of record keeping. The heat was intense and unprecedented. France averaged temperatures 10 degrees F above normal. Switzerland recorded a record high temperature of 107 de-grees F, and Britain, accustomed to cool summers and frequent show-ers, reached triple digits for the first time ever. One must remember that air conditioning was a rarity throughout much of Europe at this time. For example, in 2004, only 2,500 homes in the United Kingdom had central air conditioning. (3) During a particularly brutal two-week period in August of 2003, much of Europe experi-enced daily high temperatures in excess of 100 degrees F, and over-night lows were close to 80. A full-blown public health emergency soon followed.

Heat-related sickness escalated and exceeded the coping capacity of many European hospitals. The elderly were, as usual, the most gravely impacted. Many of them, living in cities and despite the heat and the lack of air conditioning, refused to open windows at night for fear of crime. Soon the death toll mounted, but it was not the elderly alone who made up the number of heat-related fatalities. And the numbers were staggering. On August 15, 2,200 heat-related deaths were recorded in France alone. During the course of the 2003 heat wave, over 70,000 heat-related deaths were recorded in the European Union. This compares to 1,800 deaths associated with Hurricane Katrina, for example, and places the human cost of this "heat disaster" in stark perspective. (3)

In addition to the human costs, the 2003 European heat wave can be measured in other terms as well. As a result of the heat, more than 25,000 wildfires consumed almost 1.5 million acres (600,000 hectares) of land across Portugal, Spain, France, Italy, Austria, Finland, Denmark, and Ireland. Italy saw a 40–50 percent drop in its olive production and a 40–100 percent drop in peach, apricot, and grape yields. Great Britain saw the cost of produce on the shelf increase by 40 percent. All told, economic losses from the 2003 European heat wave were conservatively estimated at around $13 billion. (3) Weather extremes of all kinds—and more of these are being both experienced and projected as the climate warms—are associated with increases in human suffering and economic harm. But they have to happen to us directly, where we live, before we seem to notice that.

Everyone familiar with the literature in the field of emergency management, for example, knows that most people do not think about disasters, natural or any other, until they happen. To be even more precise, they do not think about disasters until they happen *to them*. Even if they admit the potential for a disaster to occur, many take refuge in the thought that it will happen to somebody else but not to them. Increasingly, it is happening to all of us, and we are all feeling it. Given this general tendency to not think about disasters until they occur, it is all the more unfortunate that we must personally feel the impacts of climate change before we act to manage its threats. Climate change is one case, as we shall see as our discussion proceeds in subsequent chapters, where it may be much too late to address the crisis if we wait until we feel it in the most dramatic and persistent ways.

The United States is feeling the heat more and more. Gradually, the climate is producing human and economic impacts where we live, and it is attracting our sporadic attention. In 2012 alone, for example, dramatic examples of these impacts were impossible to miss. Out-of-season tornadoes in the Southeast and Texas, summer drought, larger wildfires, and rare tropical events in the American Northeast added up. According to the National Oceanic and Atmospheric Administration, there were 11 extreme weather- or climate-related events in 2012 that caused damages exceeding $1 billion each. Seven of these events were triggered by severe weather or tornadoes, two by hurricanes, and two by extreme drought conditions. (4) The single event that captured the most attention in 2012 and that brought the human and economic impacts to bear on more people directly where they live was probably Superstorm Sandy.

Sandy began in the Caribbean on October 19, 2012. It began as a tropical depression and became a tropical storm within six hours. It was upgraded to hurricane status on October 24. As it tore through the Caribbean, Sandy caused over $300 million in damages. Sandy then swept across the Bahamas, briefly weakened to a tropical storm on October 27th, and then strengthened again to become a Category I hurricane as it turned northward toward the United States. Sandy made landfall in the United States on the evening of October 29 near Atlantic City, New Jersey. As it inflicted its damage on the U.S. Northeast, Sandy resulted in 253 deaths and caused damages estimated at over $65 billion. The impact of Sandy on the densely populated American Northeast, its size and complexity, and the costs associated with respect to damages and losses, dominated the American media and made the discussion of natural disasters and climate-change impacts very real and very important to many people right where they live. (5)

In the immediate aftermath of Sandy, the anticipation of future "Sandys" resonated with many citizens and, surprisingly, given their usual reluctance to engage climate-related concerns, even some policy makers. Much discussion in the aftermath of Sandy included a heightened seriousness about the need to rebuild devastated areas with smarter design. The talk was not just about rebuilding, but rebuilding smarter in anticipation of future disasters of similar magnitude. The concern that Sandy may represent a "new normal" for which we must be prepared gained some traction in the immediate aftermath of its impact. While the national political debate about global warming continues, it does seem that the more we directly experience its impacts, the more we feel the heat, the more we seem willing to accept the judgment of those in the scientific community that the established risks associated with it call for an urgent and high priority response. This is a "good news–bad news" sort of deal.

The good news is that when the human and economic impacts of a warming climate are visited to our doorstep, where we live, we take note and see the need for sensible action. The bad news is double barreled. First, if these impacts do not visit our neighborhood and affect us directly, we tend to have less urgency and, in some cases, even doubt the existence of a threat or the need to act with any sense of urgency in addressing it. Second, even where the impacts have visited us directly, as time passes and we pick up all the pieces and survive, etc., we tend to feel less urgency about the need to act. It seems we require repeated and fresh experiences. It is apparent that we need to keep feeling the heat, or we easily and swiftly revert to old and often destructive habits.

While it is debated by both skeptics and deniers, we actually do know (the science is conclusive) that the observed patterns of climate change over the past fifty years cannot be explained by natural factors alone. The human production of greenhouse gases, for example, has been proven to be the major contributing factor. (6) We know that an increase in greenhouse gas concentrations in the earth's atmosphere leads to an increase in the earth's average temperature. Increasing temperatures influence the patterns and amounts of rain, reduce ice and snow cover, reduce permafrost, contribute to rising sea levels, and increase the acidity of oceans. These changes, in turn, impact our food supply, water resources, infrastructure, and our health. All of this is to say that climate change, as we are already in fact measuring and experiencing it, and as we may anticipate some of its predictable future impacts, poses identifiable short- and long-term challenges we really should respond to if we are at all logical about it.

A brief rundown of the list of things that we can say we know about the risks associated with the impact of climate change or global warming on our coasts, water resources, agriculture, ecosystems, transportation systems, and other human and natural systems suggests we will all be feeling the heat, so to speak, for a good while to come. According a recent summary of the scientific research on the subject published by the Research Council of the National Academies, future projections include more intense, more frequent, and longer-lasting heat waves. (7) But there is much more in store for us in the United States and around the globe as the climate continues to warm.

The warming of the ocean's waters and continued and more-rapid glacial melting means global sea levels will continue to rise, and with rising sea levels will come more beach erosion, loss of wetlands, increased vulnerability in coastal regions to storm surge and flooding, and possibly the submerging of coastal areas and islands. The Southwestern United States will experience continued drying that will tax water resources and expand deserts. Elevated CO_2 levels, increased temperatures, and changes in precipitation will alter agricultural growing seasons. They will also lead to an increase in insect pests, weeds, and plant diseases. Increases in wildfires across the Western United States are to be expected. Of course, there will also be an increase in public health risks from things such as heat stress, elevated ozone air pollution, food- and water-borne diseases, insects, and from direct injuries sustained in increasingly frequent and more severe natural-weather-related disasters. (7) This is but a partial list of impacts not only projected but already being felt to one degree or another in most of the places where we live. Some researchers believe that global warming is already responsible for some 150,000 human deaths per year around the globe. They say this number could more than double by the year 2030. (8)

Global warming will affect specific regions of the world and segments of society differently, of course. The physical and economic impacts will be unevenly distributed across regions and populations. And truth be told, there are things that can be done to reduce negative impacts in some regions if the people living there are smart and if they are proactive. For example, coastal cities that institute measures to protect critical infrastructure may be less vulnerable to the adverse impacts of sea level rise and storm surges. Complicating our perceptions will be the fact that climate change will make some places warmer, some cooler, some wetter, and some dryer. This variation in impacts has caused some to argue that climate change is a better term to use than global warming to convey to the public the nature of what is occurring. But it is the warming of average global temperatures that is the cause of the climate change we are already experiencing and will continue to experience. Further, and we shall see this in Chapter 3, the use of climate change versus global warming really has minimal impact on public perceptions. Global warming, with its varied impacts, is the nature of the climate change we are presently and, for the rest of our lives will be, dealing with. In the end, the terms matter less than the impacts we must understand and be prepared to manage. Thus, I tend to use the terms interchangeably and without apology.

There are some uncertainties about exactly what will happen when. It is not possible to predict the timing and the location of each specific global warming impact. However, scientific projections about the climate made over the past 50 years have proven to be incredibly accurate. Where they have erred, it has been mainly on the side of underestimating the impacts of a warming climate. There are of course no absolute assurances that projections about the next 50 years or 100 years will be equally accurate, but there is very little reason to doubt that they are based on sound science. Absolute certainty about the future is impossible. It is as impossible in science as it is in life generally. But there is virtually no uncertainty that the impact of global warming is already being felt and that it will continue to be felt over the course of the lifetime of every living being on the planet today. Even if you are among those who choose against all evidence to believe that climate change is a hoax based on junk science and promoted by Al Gore and some of his European buddies, you are feeling and will continue to feel the heat.

On a practical level, given even the most cautious interpretation of what science is able to tell us, communities around the world must be prepared to adapt to more frequent and quite probably more destructive natural-disaster events associated with the already observable and the reasonably projected impacts of a warming or changing climate. This practical conclusion gave rise to an argument I have made in some of my previous work on climate change. I have noted that preparing for the possible impacts of climate change has much in common with what we already do to prepare for, respond to, and adapt to regularly recurring and expected natural disasters. (9) In other words, just as we prepare for natural disasters, we should be able to prepare for the impacts of climate change. Indeed, I and others have suggested that the natural-hazards literature and experiences could be heavily drawn upon to guide governmental and community planning with regard to climate-change preparedness. This suggestion makes a great deal of sense. Every new natural disaster event brings with it, or so it may be suggested, a wealth of new information to consider in relation to the warming of the globe. (10) The extraordinary heat events that continue to be experienced must be mined for whatever they can teach us. There will, with absolute certainty, be a growing number of such events.

It is a summer Saturday in Las Vegas, Nevada. An elderly man is found dead in his home, a home which did not have air conditioning. Temperatures had reached the 115 degree F mark for the second consecutive day in Las Vegas. Paramedics in Las Vegas and throughout the Western United States were dealing with excessive heat and heat-related illnesses or health emergencies. (11) It was June and July 2013. Temperatures in western states were setting new records. They reached 122 F in Palm Springs, California, and exceeded 130 F in Death Valley. Civic and emergency officials throughout the Southwestern United States said, in effect, if ever there was a time to worry about a heat emergency, this was it. The following is typical of National Weather Service alerts issued at this time:

Temperatures will continue to approach or exceed records underneath an expansive upper ridge anchored over the Southwest and Great Basin. Excessive heat warnings remain in effect for a large portion of California . . . Nevada . . . and Arizona . . .

where daytime highs will yet again dangerously soar well past the century mark and overnight lows will barely drop into the seventies and eighties. As the ridge begins to build northward late Sunday and into Monday . . . triple digit temperatures will expand north through the Intermountain West and all the way to the Canadian border. (11)

Five states (California, Arizona, Nevada, Utah, and Colorado) were on "excessive heat" alerts, and heat advisories were issued for four others (Oregon, Idaho, Washington, and Montana). I'm sure that many will say, "Well, it gets hot in Vegas or in Palm Springs. Isn't that natural? What's the big deal?" The big deal is the duration and severity of the heat wave, the underlying drought conditions, and the unfolding impacts caused by it.

The summer 2013 Western heat wave kicked off a particularly deadly wildfire season in the United States. Wildfires burn about 4.3 million acres each year in the United States. In 2012, 9.3 million acres were burned. The 2013 wildfire season began with a devastating blaze on June 17, covering more than 16,000 acres in Colorado Springs, Colorado. Two people died in the fire, which also burned down more than 500 homes and forced 40,000 people to evacuate. This was followed by a wildfire in Yarnell, Arizona, in which 19 members of the elite Prescott Fire Department Granite Mountain Hotshots were killed battling the blaze. (12) The blaze from this fire grew from 200 acres to about 2,000 in a matter of hours. Tragically, the fire's erratic nature simply overwhelmed fire fighters.

It was not just the Western United States that was feeling the heat in 2013. Australia experienced a record heat wave as well. Record daily average temperatures were set over the whole of Australia. In addition to record heat, Australian newspapers warned of "Armageddon" as the spread of bushfires reached catastrophic levels. The impact was heightened by a buildup of fuel from two cooler years of unusually heavy rain. (13) For climate-change watchers in Australia, it is more the frequency of these events than the severity of any particular event that matters. The prospect of more frequent and more widespread heat waves has become a major concern. In fact, Australia's Climate Commission has said Australia can expect to experience an increase in summer heat waves that will pose increased danger to agriculture, health, and even life. A commission paper on the health impacts of climate change says there has been an increase in hot days and nights and a decrease in cold days and nights across

Australia. This study concluded that, in the past five decades, the number of record hot days has more than doubled. Concerns were elevated significantly as the recent heat waves had increased hospital admissions for kidney disease, acute renal failure, and heart attacks. Australia was expecting to continue feeling the heat for a good time to come. (13)

Nobody can say with absolute certainty, at least for several decades, that climate change is the cause of specific heat waves. Even if there were no anthropogenic changes in climate, no such thing as global warming, a wide variety of natural weather and climate extremes would still occur. This includes heat waves. This is why longer-term assessments must be made to establish climatological patterns and to distinguish them from normal short-term and natural variations. But the observation and careful analysis of the increase in the number and the length and the severity of heat waves across the globe experienced over the preceding decades does produce enough evidence to say with growing confidence the probability of a climate-change driver in relation to these events is extremely high. And there is absolutely no doubting that a warming climate does increase the number and the severity of wildfires.

How does a warming climate impact wildfires? The climate-related variables that affect the severity of a fire season are those that affect soil moisture content, vegetation density, and the moisture content of live vegetation. Extended periods of above-normal temperatures and below-normal rainfall are key factors that contribute to a more active wildfire season. Available moisture is rapidly lost due to high evapo-transpiration rates under hot and dry conditions. If the losses due to evapotranspiration are not replaced through precipitation, below-normal soil and vegetation moisture levels increase the potential for wildfire development. A warming climate and more severe wildfires are clearly linked! While it is difficult to establish direct linkages in the short-term, there is a growing body of very reliable longer-term evidence to support the linkage of climate change, or global warming, over the past 50 years, not only to wildfires but to more intense weather of all sorts. Science, which is ongoing and not without some uncertainties, is in fact telling us enough to command our attention.

The year 2014 saw a continuation of the record heat in Australia and in the Western United States. The state of California has been in a state of almost endless drought since 2011. But this situation worsened significantly in 2014, as the recurring heat waves shifted into one continuous streak of record-high temperatures that hasn't let up.

This also contributed to an increasing number of severe wildfires as the state baked and led to increased concerns about water supplies. As late as October, Los Angeles experienced record-breaking heat that forced school closures. By the beginning of 2015, water rationing in California became a necessity. (14) Yes, heat waves and wildfires do occur in the natural course of events. But as we shall see, the number and the acceleration of such increasingly severe heat events is well beyond what might be considered normal. The same can be said for a variety of other weather extremes being experienced around the globe as well. The changes in the patterns of natural disasters and the changing risks and vulnerabilities this implies for human communities everywhere should alert us to something that exceeds the normal and expected variations in climate. In fact, it is very likely that we are dealing with a new normal and that a new perspective may be needed to help us understand exactly what that might mean. It is the need for that perspective that has pretty much motivated this book.

The Perspective

My studies in the field of emergency management have convinced me that emergency management professionals are, in fact, already dealing with the impacts of global warming. It is influencing and altering the dynamics (frequency, intensity, severity, etc.) of the regularly recurring natural-disaster events that they must deal with on a routine basis. Record heat waves, more frequent and intense wildfires, the frequency and the intensity of storms, and the escalating costs of damages and disaster recovery are conspiring already to increase the heat on emergency planners and responders. Resources for disaster response are struggling to keep pace with events. The billions and ever more billions of dollars required for disaster recovery are less and less sustainable burdens for our society to bear over time. With respect to the emergency management function, the implications of a changing climate for pre-disaster planning, disaster response, disaster recovery, and pre- and post-disaster hazard mitigation are all impacts already being visited upon the emergency management professionals in the communities where they live and work.

To the degree that climate change or global warming is a long-term variable that will influence regularly recurring natural disasters, making the extreme and unlikely a new normal perhaps, it will change the risk and the vulnerability profiles of the impacted communities. As climate change continues to alter the frequency, intensity, or the nature of events, coping and response mechanisms and all natural-disaster-related planning based on past experiences will be insufficient to anticipate current and future vulnerabilities. Global warming is increasing vulnerabilities and risks, and quite dramatically in some cases. In no small part, it can be said, as I have argued in my previous work, the strategies needed to adapt to a changing climate will require new thinking and new approaches. Disaster mitigation, for example, will be more urgently required in relation to new threats and vulnerabilities. This will need to include not only the enhancement of environmental quality but, given the nature of the threats posed and more extreme events already being experienced, emphasize the need to implement plans that combine disaster risk and vulnerability reduction, pre-disaster preparedness, and post-disaster recovery to broader planning for environmental sustainability in a profoundly changing climate landscape. (9) All of this suggests the need for a perspective on global warming, or the climate crisis, that places its impacts neatly into an emergency management context. At least, that is the perspective that has sharpened my own interests.

While not a book about emergency management as such, this book will be very much influenced by my work in that area of study and some of the key concepts that may be applied to the impacts of climate change. From what I would call an emergency management perspective, global warming is perhaps best viewed as a multifaceted, multi-event, high-probability, high-impact global hazard threat unfolding over an extended period of time with increasingly observable and already experienced disaster impacts. It holds the all-too-real potential for an escalating and increasingly devastating series of future disaster impacts. This is to view climate change as a crisis already in motion. The question is—and it is a question that is a large part of the motivation behind this book—is global warming a climate crisis that can be managed? The effort herein to answer that question will be influenced in large part by my emergency management perspective. Such a perspective, or so I would argue, holds the potential to constructively focus the climate change or global warming conversation on the practical impacts that must be objectively anticipated and managed intelligently.

An emergency management perspective requires first that one understand what is meant by the words "climate crisis." The implication is, of course, that global warming (whatever its causes) poses serious threats and that we and the communities in which we live and work are vulnerable in the shadow of those threats. The threats are escalating, and our vulnerabilities are changing. Thus we must, as a practical matter, begin with a reassessment of threats and vulnerabilities. How are we feeling the heat? What will the future entail for us? Secondly, we must discuss the strategies available to us to adapt to a warming climate and/or to mitigate the worst or most dramatic of its potential impacts. Thirdly, we must understand that we need to adapt to survive the impacts we cannot prevent. There will undoubtedly be many more of these unpreventable impacts than we are able to predict. For all that we may know and that science may enable us to anticipate with reasonable assuredness, there is also a very real possibility of impacts that have not yet been anticipated. The situation is fluid, and what we do or refuse to do in the short term may alter the course of future events significantly.

Before following the logic of what I am calling my emergency management perspective, a perspective that will begin with the assessment of risks and vulnerabilities and logically flow to the identification of reasonable mitigation and adaptation measures that can reduce them, there are a few other foundational bases that absolutely need to be covered. These bases must be covered in order to understand what is being referred to as the climate crisis. First, we must have an overview or foundation for understanding exactly what science is telling us about climate change. I am not a climate scientist doing climate research, of course, but relying on the best available peer-reviewed scientific research, a reliable summary of what science is telling us can be drawn. In providing that basic summary, all efforts will be made to be accurate in providing the broad overview of what science can tell us about global warming. As we examine the scientific portrait, it will be essential to evaluate the arguments of global warming deniers and skeptics. Their arguments cannot be ignored as a part of the ongoing public discourse about the subject. To make the case that there is in fact a climate crisis, their arguments must be reasonably addressed and put into scientific perspective.

The politics of and the public's knowledge about climate change in the United States must also be discussed and understood, for that too is a crucial part of the story, and it has a very direct bearing on the climate crisis and our ability to define and manage it. Climate science, as we shall see, is limited in its ability to influence public policy and public strategies that respond to the "problem." Our politics often work to impede our progress in moving forward with mitigation and adaptation efforts. That must be viewed as a part of the "problem" or "crisis" if we are to obtain a full understanding of it. In other words, the climate crisis is made worse by a political crisis. Both must be managed more intelligently. This discussion must also assess the role of public opinion and the role of citizen involvement in the efforts to define and manage what we are calling the climate crisis. Indeed, as a discussion of the politics of climate change may reveal, an informed and involved citizenry may be one of the most important variables to be cultivated and enhanced in our efforts to manage the climate crisis.

Once the preliminary bases have been covered (the science, the deniers/skeptics, the politics, and the public role), we will have a fuller context for the application of some of the basic emergency management thinking and logic. This will be aimed at a discussion of climate change mitigation and adaptation strategies and some of the central concerns that have influenced my own interest in relation to climate change and its impacts. This will include an assessment of threats and vulnerabilities, a discussion of adaptation and mitigation options, a discussion of adaptation strategies, and some important implications for public policy. One of the merits of this sort of analysis, focusing on identifiable risks, vulnerabilities, and logical mitigation and adaptation measures to reduce them, might be that it will enable policy makers and citizens generally to speak the same language, participate in the same analysis, and pursue the same goals. Is this too idealistic? Probably it is. But if we can in fact begin to agree that we have a crisis to manage, perhaps progress along these lines is possible. For emergency management professionals accustomed to this sort of thinking, for the communities they serve, for the interested and concerned citizen feeling the heat, and even for the skeptic, this book may provide a perspective that enhances communication and the prospects for thoughtful cooperation. The deniers of climate science will not be reachable in such an effort, but such an effort by the rest of us may reduce their number. Reducing the number of deniers is, one might suggest, a critical part of the needed response to the crisis.

As we prepare to move this discussion forward, I cannot help but reflect on some of the "discussion" in the immediate aftermath of Superstorm Sandy in the fall of 2012. The question of whether this extraordinary event was caused by or was evidence of global warming was widely discussed, and without universal agreement, it must be added. In all honesty, one must note that no single storm can be attributed to global warming in the immediacy of its occurrence. So any such discussion must be tentative and incomplete, whatever argument one wishes to make.

Some argued that Sandy was a freak combination of unlikely but not impossible events that created a "perfect storm," and that none of this had anything to do with global warming. Some, including the governor of New York and other elected officials, as I recall, stated publicly that Sandy was proof that we need to take climate change seriously. One of the unavoidable hazards of the business of assessing such things in the immediate aftermath of a disaster occurrence is that such debate occurs and opinions are formed long before the science is done to support any particular conclusion. It is important to note that science can be very deliberate, and 100 percent certainty is usually unattainable. But science, of course, is not about certainty. Science is about evidence, and its work is forever ongoing. Gathering the evidence and subjecting it to legitimate scientific analysis does give us the basis for reaching conclusions, and the reliability of such conclusions can be verified by further observation and testing. Science, as such, is all about the inquiry, and its search for refined understanding and enhanced explanation is never ending. But it does, at its best, reach a consensus that provides us with the most reliable and the very best attainable foundation for what we might call emerging scientific truth.

In the spring of 2013, Cornell and Rutgers researchers said their scientific assessment produced evidence that Sandy was made far worse by the melting of Arctic sea ice. This, they said, intensifies air mass invasions toward the middle latitudes, thereby increasing the frequency of atmospheric blocking events like the one that apparently pushed Hurricane Sandy west into New Jersey and New York. They concluded that

> If one accepts this evidence and . . . takes into account the record loss of Arctic sea ice this past September, then perhaps the likelihood of greenhouse warming playing a significant role in Sandy's evolution as an extra-tropical super-storm is at least as plausible as the idea that this storm was simply a freak of nature. (10)

Now, to be perfectly honest, in the eyes of many global warming deniers, words like *perhaps* and *plausible* are reasons in and of themselves to dismiss this study. But to dismiss the science in this case, which the deniers do, is to misunderstand what science is all about. To say it is plausible, based on scientific evidence, is not the end of discussion. Likewise, we cannot say that because it is plausible but not certain that these scientists are not to be believed. This study is a beginning. Scientific analysis will continue. The point is not to get so caught up in either the plausibility or the uncertainty of any preliminary study so as to accept or reject its conclusions as final. Science is an ongoing analytical activity, not an assertion of undeniable truth. It is an ongoing quest for more precision and more confirming or disconfirming data. It constantly tests and questions itself with the perfectly rational awareness that scientific work is always incomplete. Its work is never done!

One must be careful to avoid saying that because the science is always incomplete, and hence it is never 100 percent certain, that this means that the science is faulty or, in some more non-scientific and everyday language, a hoax. Newsflash! If it's a hoax, it isn't science. Likewise, if someone is absolutely certain about anything, their conclusion is generally not scientific. Absolute certainty is, in fact, one of the greatest of all hoaxes. The point is, once again, that science does not produce certainty. Scientific research produces, as we have already said, evidence. If, over time, repeated study and testing by numerous scientists confirms the evidence, then there is something called "accepted scientific knowledge." This does not mean that all questions have been answered, and scientists will continue to address those that are unanswered in their ongoing work. What it means is that there is sufficient evidence that has met all of the tests for reliability, and there is a consensus of expert opinion about the matter. That is exactly what scientific knowledge is—a consensus of expert opinion. A single expert, or a small handful of dissenting experts, may disagree with the accepted knowledge, but they are not presumed to represent a proof that the consensus of expert opinion is flawed or wrong. Their task, if they wish to challenge the consensus of expert opinion and ultimately improve or change it, is to produce the scientific work and uncover evidence that may alter or change the consensus among researchers active in the field. So far that has not happened with regard to the consensus among climate scientists about global warming.

The evidence that has produced the accepted scientific knowledge (consensus) on global warming, as we shall review it in Chapter 2, is compelling. It will be argued that this consensus alerts us to a crisis that makes the question of how we will adapt to and mitigate more extreme heat waves, more severe storms, coastal erosion, floods, and other impacts associated with global warming a very real one for every community where we live. This question makes climate change a profound concern for every community around the globe. Ideally, it would be better to reach an agreement on what sort of challenge or crisis we are facing and to begin acting in response to it before we feel the worst of the heat. And we do continue to feel the heat.

The reliably predictable impacts of a changing climate on communities, economies, human populations, and on the environment generally suggest that there is an already occurring and growing crisis to be managed. This crisis may be understood, for someone like me who studies natural disasters, as an emergency management challenge and a priority for those who, in particular, must take the lead in reducing the risks of future disasters. What we can safely say we know about global warming, all the while admitting that there are of course things we do not know, should tell us that we do not have the luxury of waiting to see what happens before we begin to prepare for its impacts. It is, as we shall see, too late to stop global warming from negatively impacting our lives and pointless to deny that it will continue to do so. Its impacts are already being felt. Its future impacts will be inevitable and ongoing for some time. But we can work to avoid the worst of its future effects, identify their risks, develop strategies to reduce those risks, and make the problems associated with the climate challenge smaller for our children and grandchildren. We can, if we should choose to do it, manage the crisis.

Before we will be able to talk of crisis management, before we can begin to make sense of such a conversation, we must see what science is able to tell us about the nature of the challenge posed by global warming. In the process of that discussion, we can better define the challenge as a crisis. This is not a crisis far off in our future, but one already in motion. We are not only feeling the heat, we are also presented with the evidence that should compel us to act.

References

1. "NCDC Announces Warmest Year on Record for Contiguous U.S." National Climatic Data Center. http://www.ncdc.noaa.gov/news/ncdc -announces-warmest-year-record-contiguous-us (accessed May 21, 2013).

2. Barnosky, A.D., Hadly, E.A., Bascompte, J., Berlow, E.L., Brown, J.H., Fortelius, M., Getz, W.M., Harte, J., Marquet, P.A., Martinez, N.D., Mooers, A., Roopnarine, P., Vermeij, G., Williams, J.W., Gillespie, R., Kitzes, J., Marshall, C., Matzke, N., Mindell, D.P., Revilla, E., and Smith, A.B. (2012). "Approaching a State Shift in Earth's Biosphere." *Nature* 486, 52–58.

3. Stone, Brian Jr. (2012). *The City and the Coming Climate*. Cambridge, Cambridge University Press.

4. National Oceanic and Atmospheric Administration. http://www.ncdc.noaa.gov/billions/ (accessed August 24, 2015).

5. "Hurricane/Superstorm Sandy October 2012." http://www.hurricanes-blizzards-noreasters.com/sandy.html (accessed May 21, 2013).

6. Wigley, T.M.L., Ramaswamy, V., Christy, J.R., Lanzante, J.R., Mears, C.A., Santer, B.D., and Folland, C.K. (2006). Executive Summary. In Karl, T.R., Hassol, S.J., Miller, C.D., and Murray, W.L. (Eds.) "Temperature Trends in the Lower Atmosphere: Steps for Understanding and Reconciling Differences." *Synthesis and Assessment Product 1.1 U.S. Climate Change Science Program*, Washington, DC, 1–15.

7. National Research Council (2011). *America's Climate Choices*. Washington, D.C., The National Academies Press.

8. "The Impact of Global Warming on Human Fatality Rates." *Scientific American* June 17, 2009.

9. Schneider, Robert O. (2013). *Emergency Management and Sustainability: Defining a Profession*. Springfield, Il., Charles C. Thomas Ltd.

10. Taylor, Kate. (March 6, 2013). "Global Warming Turned Sandy into Superstorm." http://www.tgdaily.com/sustainability-brief/69956-global-warming-turned-sandy-into-superstorm (accessed August 24, 2015).

11. Memmott, M. (2013). "Western States Heat Wave Turns Deadly; No Relief in Sight." *National Public Radio*. http://www.npr.org/blogs/thetwo-way/2013/06/30/197229659/western-states-heat-wave-turns-deadly-no-relief-in-sight (accessed July 10, 2013).

12. Center for Disaster Philanthropy. http://disasterphilanthropy.org/disasters/western-wildfires-2013/ (accessed July 10, 2013).

13. The Australian News (2013). The Great Heat Wave of 2013. http://www.theaustralian.com.au/news/features/the-great-heatwave-of-2013/story-e6frg6z6-1226549810192 (accessed July 10, 2013).

14. Spross, J. (2014). "There's A 99 Percent Chance 2014 Will Be the Hottest Year California Has Ever Seen." *Climate Progress*, November 6, 2014. http://thinkprogress.org/climate/2014/11/06/3589752/california-expected-hottest-year-2014/ (accessed August 24, 2015).

CHAPTER 2

What Does the Science Tell Us?

Introduction

Climate change is not a recent or new phenomenon. The science surrounding it is well established, and climatologists have been studying it for over 150 years. The climate is constantly changing due to natural phenomena and, as we shall see, due to human activity as well. Climate change is not necessarily a problem or, as we are suggesting about the current situation, a crisis. Not all natural or anthropogenic (human caused) influences on the climate produce a serious threat. But any climate changes that exceed the norm, that may test the resiliency of natural ecosystems, can pose difficulties that logically must be noted and addressed. Whether the causes of the difficulty are natural or of human origin, extreme shifts in climate may have negative impacts on all life on the planet. The present scientific consensus surrounding the phenomenon of global warming warns us of serious problems about which we should be concerned and constructively engaged.

The *climate* of the earth, by which is meant the average of meteorological conditions over time, does change in the natural course of events. Scientists have been studying the ways the climate changes, both in particular places and for the earth as a whole, since the beginning of the 19th century. This includes the most basic of questions. Why is the earth the temperature that it is? What accounts for changes in average temperature or changes in other meteorological measures such as humidity, wind speed, cloud cover, precipitation, and snow and ice cover? What difference does all of this make for life on this planet? The climate impacts people directly, as in the case of damaging storms or deadly heat waves, but it also impacts all environmental and ecological processes. Many of these impacts or effects may not be immediately recognizable to us, but they do hold the potential over time to alter our lives significantly. Some of these alterations can be very negative indeed. Large and exponentially accelerating changes in climate may, for example, pose threats to the air we breathe and the water we drink. Such changes may endanger ecosystems and threaten coastal zones, wetlands, and the stratospheric ozone layer. Thus, it can be said that studying and understanding changes in our climate over time is important and is connected to our social and personal well-being. Much of our human existence is related to climate. What we build, how we live, how we earn our livings, what we do for recreation, the food we eat, the air we breathe, our health, and much more depends on a benign and narrow range of climatic conditions.

One particular aspect of climate makes it very difficult for us to factor it into our thinking and the decisions we must make on personal, social, and political levels. Climate is a long-range proposition. As such, the long-haul trends matter more than the conditions we are experiencing at any particular moment. We do well with short-term matters such as weather. We watch the weather and, based on the immediate forecast, make day-to-day decisions quite rationally. We decide to carry an umbrella if rain is predicted, we decide if this is a good weekend to go fishing, or we decide to protect our plants against the frost, etc. Climate, or meteorological conditions over time, is connected to longer-term decisions about more complex concerns. Changes in average temperatures or levels of precipitation over time may have implications for food supply, electrical demand, water supply, and public health, to name just a few. But because the impact is spread over a longer time frame and is not always immediately recognizable to us, climate change is much more difficult for us to think about or manage rationally than our day-to-day responses to the weather.

The purpose of this chapter is to provide some of the necessary scientific background to explain the basics of climate change and to provide a context for understanding the assertion that, due to global warming, a climate crisis is already at hand. There is, one might agree, room for honest and informed disagreement over what we should do about global warming. But the state of scientific knowledge about climate change is compelling enough to make understanding the likely impacts of global warming and being prepared to deal with them a priority. The point to be made throughout the discussion in this chapter is that anticipating and responding to the challenges posed by the already measurable and projected changes in our climate is an immediate and practical necessity. The point of view here is that the impacts already being felt and those reasonably projected for the future suggest quite convincingly that a crucial or decisive moment is at hand, a turning point that calls for a change of direction if we are to avoid the most severe impacts of a changing climate. That is what is meant by use of the term *climate crisis*. There are impending hazard risks, and we are incredibly vulnerable to them. These risks are changing and multiplying. The science is sufficient, whatever its uncertainties or unresolved questions, and whatever our doubts about it, to demonstrate that our present course is unstable and unsustainable. We need to change course. *We must either manage the crisis or be managed by it.* Those are the only options we have. It matters little if we see it coming or not, because the crisis is actually already here. But before this discussion may even begin, we need to have some sense of the science that informs this view.

The Warming of the Earth's Atmosphere

Before we can speak of managing a crisis, we must first understand what it is. A quick review of some basics is necessary and may help us understand the contemporary discussion about climate change (a.k.a. global warming). The source of energy for the earth's climate is, of course, the sun. The sun emits photons (tiny packages of electromagnetic energy) of varying wavelengths. These photons interact with the matter of the world. We are actually able to see some electromagnetic radiation in the form of color, the size of the wavelength altering the perceived color. Photons of longer wavelengths, called infrared, are not visible to our eyes. All of the objects we see are constantly absorbing photons and their bits of energy. These worldly objects also reflect or emit photons and absorb the photons reflected by other objects. If an object releases more energy by emitting more protons than it absorbs, it will cool. Conversely, if an object gains more energy by absorbing more photons than it emits, it will warm. (1)

Almost all photons that strike the earth come from the sun. This fact gave rise to an interesting question first asked by a 19th-century French mathematician, Jean-Baptiste-Joseph Fourier. (2) If the earth is constantly absorbing energy from the sun, why does it not heat up until it is as warm as the sun? The answer to this question is that the earth and its atmosphere reflects or radiates some energy back out into space, thereby offsetting energy absorbed from the sun. But it does not reflect all of the energy back into space. If it did, the planet would be much too cold for human life. Fourier reasoned that some heat had to be retained in the atmosphere. It does not all reflect back into space, because some of it is absorbed (trapped) by atmospheric gases. In other words, incoming solar radiation (short wavelengths) passes freely through the earth's atmosphere to the earth's surface, but some outgoing terrestrial radiation (emitted on a longer wavelength) is trapped by the earth's atmosphere. The gases in the earth's atmosphere act as a blanket insulating the planet, thus raising the surface temperatures of the planet. This is much the same as when you cover up with a blanket. It traps heat, and you warm up. Remove the blanket, and you cool down. It is in just this sense that some of the earth's gases act as a blanket. This is what we call the greenhouse effect. (2)

The atmosphere must absorb more radiation than it reflects or emits to produce a warming effect. Many gases exhibit *greenhouse* or blanketing properties. Some of them occur in nature (water vapor, carbon dioxide, methane, and nitrous oxide), while others are exclusively human made (like gases used for aerosols). Water vapor (H_2O) is responsible for approximately two-thirds of the greenhouse effect. The largest contributor to the greenhouse effect after water vapor is carbon dioxide (CO_2) followed by methane (CH_4). (1) As the quantity of greenhouse gases in the atmosphere increases, less energy is reflected back into space, and more is trapped on the planet. The blanket gets thicker, we might say. This produces an increase in the warming of the planet's surface. The natural interaction of the earth's atmosphere with incoming solar radiation, that is, the trapping of reflected terrestrial radiation, increases temperatures enough (about 60 degrees F) to enable the planet to sustain life. Lacking an atmosphere, lacking any insulating or blanketing effect, the earth would be dark and cold like the moon. It would not sustain human life. The first direct measurements of the absorptive capacities of the earth's atmosphere were provided by the experimental physicist John Tyndall in 1859.

Tyndall constructed the world's first ratio spectrophotometer. This is an instrument used to determine the intensity of various wavelengths in a spectrum of light. He set about to test Fourier's hypothesis regarding the absorptive capacities of the earth's gases. Tyndall was actually able to determine the degree to which specific gases present in the earth's atmosphere were able to absorb or transmit radiation. He found that the most common gases, oxygen and nitrogen, were transparent to both solar (short-wave) and terrestrial (long-wave) radiation. But other gases, such as water vapor, CO_2, and CH_4 were transparent to radiation emitted by the sun (incoming) but opaque to radiation emitted by the earth (outgoing). In other words, they reacted differently or selectively and, with respect to terrestrial radiation, some heat is trapped in the earth's atmosphere by the blanketing effect of CO_2 and other greenhouse gases. This proved that the greenhouse effect, though not yet named as such, was real. (3) Tyndall also raised the question of the potential for changing atmospheric CO_2 levels to cause measurable changes in the earth's climate. But this question would not be explored until after his death.

A Swedish physicist named Svante August Arrhenius, after Tyndall's death in 1893, was the first to detect the influence of carbonic acid in the air on ground temperatures. (4) Arrhenius's exhaustive computations, confirmed by subsequent study over the next hundred years, showed that significant changes in atmospheric CO_2 levels could indeed lead to some very profound shifts in global climate. For example, he demonstrated that a reduction of CO_2 levels by half (based on the levels at his time) would lower the average global temperature by seven to nine degrees. This would be enough to cause a new global ice age. Conversely, the doubling of these CO_2 levels would cause increases in average global temperatures of between nine to eleven degrees. This, he said, would cause glaciers to retreat and sea levels to rise, and would result in considerable global warming. Arrhenius believed that a warming trend was the most likely outcome for the future based on the pace of industrialization during his lifetime and the increasing reliance on fossil-fuel combustion for energy. He even felt that some warming was a desirable thing. This did not worry him greatly because, based on the then current rates of carbon production, he computed that it would take 3,000 years to double CO_2 levels. (4) This was, one might suggest, the one computational imperfection in his analysis. Not able to foresee or include in his calculations the implications of a rapidly industrializing society and the increasing speed at which it would facilitate the emission of billions of tons of carbon into the atmosphere, Arrhenius's estimate of the doubling of CO_2 levels was radically underestimated. Indeed, factoring in CO_2 levels at the carbon accumulation rates of the late 20th century, the time required for the doubling of CO_2 is reduced to a mere 100 years. For us, the distant future turns out not to be as distant as it was for Arrhenius. (3)

The linkage between the greenhouse effect and fossil-fuel consumption has been well established for over 100 years. This seems, sometimes, to be a matter of dispute outside of scientific circles and among deniers in the political realm, but it has long been settled science. The role of CO_2, CH_4, and other greenhouse gases in warming the climate is not a matter of scientific dispute. But what about the role human activities play in relation to the production of greenhouse gases? Are human activities enhancing the greenhouse effect in a major way? This would remain a subject of credible debate until the late 20th century. But the answer began to develop during the 1950s through the work of Charles Keeling. Keeling, a research scientist at the Scripps Institution of Oceanography in California, installed one of the first manometers (a device designed to measure atmospheric carbon dioxide) at the Mauna Loa Observatory in Hawaii. Keeling would ultimately expand his CO_2 research to diverse areas such as the Big Sur near Monterey, California; the Olympic Peninsula in Washington; and the mountains of Arizona. As his various monitoring stations have gathered data over the years, much has been learned that should inform our global warming discussion today. (5)

Keeling would be the first to discover the seasonal rhythm of CO_2 levels. In Mauna Loa in 1958, he observed that CO_2 levels peaked in May and then dropped to a yearly low in October. Repeated observation of this pattern in subsequent years led him to conclude that he was observing the withdrawing of carbon from the air for plant growth during the summer and returning of it each winter. This was a natural cycle. (5) Among the other things observed, Keeling was able to determine whether the concentration of carbon dioxide in the atmosphere was uniform across the earth. The Mauna Loa manometer would confirm that it was. This meant that higher emissions in any portion of the planet would ultimately increase CO_2 concentrations in all regions of the planet. Over time, a much more profound and indisputable discovery would be made. In addition to being globally diffuse, CO_2 levels were found to be steadily and quickly rising due primarily to human influences. (3)

Keeling would observe that the amount of CO_2 in the earth's atmosphere was steadily increasing. This, he was able to demonstrate, was connected to the combustion of fossil fuels. Moreover, the increase in CO_2 levels was sharper with each successive year. (5) As shown in Keeling's charting over more than 50 years, there has been a dramatic upward curve in carbon emissions from the middle of the 20th century onward. Known as the Keeling curve, this record is considered to be among the best and most consistent sources of atmospheric data. From the mid-1950s to the present, carbon emissions have increased from about 315 parts per million to over 400 parts per million. (3) The measurable climb in carbon dioxide started well before Keeling began watching, of course. More recent research has shown that preindustrial levels of carbon dioxide (from AD 1000–1750) ranged between 275 and 285 parts per million. (1) Most of the measured growth associated with industrial development appears to stem from human activity. That is to say, it is associated primarily with the use of fossil fuels. Today, about one in four CO_2 molecules in the atmosphere comes from us.

As shown on the Keeling curve over the past 40 to 50 years, the rate at which CO_2 levels are increasing is considered to be unprecedented in human history. According to climate researchers, the "safe" or upper level of carbon dioxide in the atmosphere should not exceed 350 parts per million. *Safe* here means conditions that will minimize the likelihood of the most severe negative effects from a significant climate change in the form of rising seas, wildfires, and extreme weather of all kinds. In early 2013, CO_2 levels reached 400 parts per million. It has remained at this level. Indeed, the Scripps Institution of Oceanography website has daily updates of emissions levels. For example, the reading for April 5, 2015, was 403.64 parts per million. The point is CO_2 accumulations have not reached this level at any previous time in human history. Scientists are concerned about the consequences this holds in store for the planet. According to the CO_2 Program at the Scripps Institution, we are on the threshold of a new era in geologic history. It will be, or so they say, one in which the climate will be very different from anything previously experienced by humanity.

Increasing levels of CO_2 in the atmosphere are a problem in general because of the long period of time required to cycle them out of the atmosphere. More than half of the carbon dioxides released each year from all sources remains in the earth's atmosphere for many decades. Twenty to 35 percent of it remains in the atmosphere for several hundred to over 1,000 years. Carbon, like other common elements, is constantly recycled on the land, in the oceans, and in the atmosphere through a natural process. But this process takes time. All of this is complicated by the fact that the rate at which fossilized carbon is being returned to the atmosphere is much greater than the rate at which natural processes can absorb them. (3)

The human contribution to the problem of increased CO_2 levels is the result of our dependence on fossil fuels for energy, which means the burning of oil, gas, and coal. During the past 20 years, about three quarters of human-made carbon dioxide emissions (and these are still increasing) were from burning fossil fuels. It can be said that we, through our energy development and consumption patterns, contribute to the creation of an imbalance in the global carbon cycle. This imbalance means that carbon is accumulating more rapidly in the atmosphere than it is being removed through absorption. This excess stock of carbon dioxide enhances the greenhouse effect and warms the climate. (3) As Keeling concluded, this translates into a growing blanket of carbon dioxide, which raises the earth's average temperatures. While doubters exist, scientists have been able to accurately reconstruct a record of carbon dioxide and temperature trends that reaches back to ancient millennia. Such reconstructions have confirmed the close association between atmospheric levels of CO_2 and temperature. Perhaps the most extreme example of this relationship may be the planet Venus. About 97 percent of the Venus atmosphere is composed of CO_2. As a result, almost all of the heat emitted from the planet's surface is trapped on the planet by the atmosphere. This contributes significantly to temperatures in excess of 800 degrees F. (3)

Scientists have shown the correlation between CO_2 and higher temperatures from 1880 to 2000. In other words, the evidence is conclusive that rising CO_2 levels are correlated with global warming. However, linked as they are, CO_2 levels and temperatures do not rise at the same rate. There is, due to the absorptive effect of the oceans for example, a lag between the two. This creates room for doubters and deniers to suggest that the linkage between rising CO_2 levels and temperatures is exaggerated or not settled, but the science has made it clear that the absorption of carbon by the oceans is what accounts for the delaying effect. Despite the need to always investigate and refine our understanding of the relationship between CO_2 and temperature, there is a clear and undeniable causal relationship between rising atmospheric CO_2 concentrations and increasing global temperatures. Arrhenius had it right all those years ago.

One suspects that the efforts to downplay or even deny the relationship between CO_2 emissions and temperature has something to do with resistance to the notion that human activity is contributing to the warming of the earth. This fact that we humans do affect or change the climate causes more consternation than it should, actually. If we concede that the reflectivity of the earth's surface and its atmosphere is not perfect, we accept that it does not send all short-wave radiation received from the sun back into space as a perfect mirror. This means that a percentage of this radiation is absorbed and returned to the atmosphere as long-wave radiation that is trapped by the gases in the atmosphere. We accept that the natural greenhouse effect that warms the planet is driven by the proportion of solar radiation absorbed by the earth's surface and converted into long-wave radiation. The earth's reflectivity, its *albedo*, is quantified as the proportion of short-wave radiation received from the sun that is reflected back out into space. Earth's natural land covers and oceans always absorb some percentage of this energy, and they return it to the atmosphere (over time) in the form of long-wave radiation. The earth's reflectivity is thus subject to change. Some of this is due to entirely natural causes, of course. For example, the classic example of a natural albedo effect is the snow-temperature feedback. If a snow-covered area warms and the snow melts, the albedo decreases. This means more sunlight is absorbed and that temperature tends to increase. The converse is also true. If snow forms, a cooling cycle happens. But human activities have also changed the albedo through activities such as fossil-fuel consumption, forest clearance, and farming around the globe. (3, 6) This is really pretty simple and straightforward.

A perfect example of the human influence on the earth's albedo, and subsequently on its climate, may be urban development. The rate at which natural land cover is modified is a contributor to changes in our climate just as is the loading of the environment with greenhouse gases that trap in the heat. In fact, landscape change in cities has been cited by some as one of the principle climate-related threats to human health today. (3) There is also evidence to suggest that the landscaping of cities may impact the climate as much or even more than the deforestation of rural areas. The displacement of vegetation by streets, parking lots, and buildings contributes to a reduction in evaporative cooling. Paving and roofing materials absorb more solar energy and return more long-wave radiation that lowers surface reflectivity. Tall buildings absorb more reflected solar energy and the heat generated by mechanical sources. Automobiles, waste, and air conditioning all generate some of the heat we feel in our cities. This is what we call the urban heat island effect.

Virtually everything we humans do, from the development of our cities to the production and consumption of energy, from the way we farm the land to the way we travel, etc., has the potential to impact climate. There is a preponderance of scientific evidence that connects human activities to most of the global warming that has occurred over the past 50 years. (7) This is especially true with respect to the release of CO_2 and other heat-trapping greenhouse gases in the atmosphere. Let us briefly consider in a little more detail what the science is able to tell us about the human contribution to the global carbon imbalance.

The concentration of CO_2 in the atmosphere has increased markedly over the past 150 years. It is now in fact higher than at any time in well over 800,000 years. (8) The observed long-term increase in CO_2 levels is directly attributable to the growth in human carbon emissions from fossil fuels. (9) Growing concentrations of other greenhouse gases (methane, nitrous oxide, etc.) are also attributable to human activities. (10) Both natural physical principles, such as those we have discussed, and scientifically reliable models of the earth's climate system show that, without a doubt, when greenhouse gas concentrations increase, planetary warming will occur. (7) Most importantly, and conclusively, careful and reliable scientific analysis demonstrates that natural factors such as normal internal climate variability or changes in energy emitted by the sun do not explain the current and projected global warming trend. (11) We will discuss this at some greater length later, but the evidence is that the global warming we are presently experiencing is not due to natural causes. And make no mistake, the climate is warming. Between 1900 and 1988, for example, new temperature heat records were established on an average of every 12.6 years. In the decade between 1988 and 1998, the interval for new heat records had been reduced to 2.5 years. All of the hottest years ever recorded by human instruments have been recorded since 2001. The statistical probability that such a string of increasingly hotter years could occur absent an underlying shift in the climate is exactly zero.

Whether or not one wishes to accept the scientific evidence that human activities are a contributing factor to global warming, there is ample evidence that the earth's climate is changing and that the changing climate can be linked to changes in the earth's temperatures. On a global scale, average temperature is warming. Among the things that can be directly linked to increasing global temperatures are the widespread melting of glaciers and ices sheets and decreases in Northern Hemisphere snow cover and Arctic ice. The permafrost is thawing, lakes and rivers are freezing later and melting earlier, and agricultural growing seasons are being altered in some locations. Of course, despite any of these measurable phenomena, people still raise the question of whether global warming is really happening and whether it is a problem that we must deal with now or at any future time. Let's see if we can summarize the scientific answer to that question.

Global Warming: Is It Really Happening?

Global climate-change impacts in the United States have been accessed and reliably summarized by the U.S. Global Change Program. (12) The U.S. Global Change Research Program (USGCRP) is a Federal program that coordinates and integrates global climate-change research across 13 government agencies. Its purpose is to quantify changes in the earth's climate systems, improve climate-change projections for the future, and advance the understanding of American policy makers and citizens of the vulnerabilities we may face in relation to climate-change impacts. The USGCRP was established by Congressional mandate in the Global Change Research Act of 1990. It has since made the world's largest scientific investment in the areas of climate science and global change research. (12) In a description of their work, the USGCRP states that "the environment is changing rapidly. Increases in world population, accompanied by industrialization and other human activities, are altering the atmosphere, ocean, land, ice cover, ecosystems, and the distribution of species over the planet." (12) The USGCRP website goes on to say that "understanding these and other global changes, including climate change, is critical to our Nation's health and economic vitality. Scientific research is critical to gaining this understanding." (12) The accompanying list provides a quick summary of recent USGCRP reports as summarized by the National Research Council of the National Academies. This provides a quick overview of some of the generally observable impacts of climate change on the United States.

Changes in Climate Observed in the United States

- U.S. air temperatures have increased by two degrees F over the past 50 years; total precipitation has increased by about five percent over the past 50 years; sea level has risen along most of the U.S. coast;
- Sea level rise is eroding shorelines, drowning wetlands, and threatening the constructed environment;

- Permafrost temperatures have increased throughout Alaska since the late 1970s causing significant damage to roadways, runways, water and sewer systems, and other infrastructure;
- There are widespread temperature-related reductions in snow-pack in the northeastern and western United States over the past 50 years, causing changes in seasonal timing of river runoff;
- Precipitation patterns have changed: heavy downpours have become more frequent and more intense;
- The frequency of drought has increased over the past 50 years in the southeastern and western United States;
- The frequency of wildfires and the length of the fire season have increased very substantially in the western United States and Alaska. (7)

Looking beyond the United States, the evidence for fairly rapid climate change on a global scale is compelling. Global temperature, according to all major surface temperature reconstructions, has warmed since the 1880s. The warmest 20 years on the planet have occurred since the 1970s with all 10 of the warmest years having occurred since 2000. (13) Since the beginning of the 2000s, we have witnessed a decline in solar output, yet surface temperatures have continued to increase. (14) The oceans have absorbed much of this increased heat, resulting in increasing water temperatures.

The shrinking of the ice sheets has been stunningly noticeable. The ice sheets in Greenland and the Antarctic have decreased in mass (36–60 cubic miles of ice per year in Greenland), and both the extent and the thickness of Arctic sea ice has declined rapidly for several decades. (15) Glaciers are in retreat, and the retreat is accelerating noticeably almost everywhere including the Alps, Himalayas, Andes, Rockies, Alaska, and Africa. (16) Extreme weather-related events are increasing rapidly as well. The number of record-high temperatures in the United States has been increasing, while the number of record-low temperatures has been decreasing since 1950. Additionally, the increasing number and increased intensity of natural disasters in the United States is linked to climate change. (17)

Among the most interesting and compelling indications of both a changing climate and an anthropogenic linkage to it is ocean acidification. Since the beginning of the Industrial Revolution, the acidity of ocean water has increased by roughly 30 percent. Studies have confirmed that this increase is the result of human activities that emit more CO_2 into the atmosphere and, subsequently, more of it is being absorbed into the oceans. Presently the amount of CO_2 absorbed by the oceans is increasing by about 2 billion tons per year. (18, 19) All of these observable global climate-change phenomena have profound implications for life on this planet. Let us examine a few of them just a little more closely.

Global average sea level has increased eight inches since 1880. In some locations, such as the Gulf of Mexico and along the U.S. East Coast, sea levels are rising much faster. Projections are for continued rises in sea level. (20) Global warming is considered to be the primary cause of current sea level rise. Heat-trapping gases have increased global temperatures. Rising temperatures cause the oceans to warm. Sea level rise is the result of seawater expansion as temperature rises and the shrinking ice sheets and glaciers add water to the world's oceans. Ice sheets, ice caps, and melting glaciers have contributed about half the total of sea level rise between 1972 and the present. (20) The rise in sea levels is accelerating.

The global sea level record from 1880 to 2013 (an eight-inch rise) is a fact. Projections for future sea level rise are, of course, estimated. This often invites skepticism. How do we know such estimates are reliable? There are upper- and lower-limits in most of the projections, depending on a range of factors and variables. The projections establish a range, but the science that these projections are based on is robust and certain that future sea level rise will be very significant and very troubling in any scenario that may come to pass. (20) Just like global surface temperatures, the rise in sea levels will be greater in some locations than others, but it is accelerating everywhere, and this holds the potential to alter the course of tropical events, increase storm surges and flooding, destroy more constructed infrastructure, and inundate islands and coastal communities. (21)

There is growing scientific evidence to link climate change to extreme weather. With respect to heat waves and coastal flooding, the scientific evidence in support of this linkage is clearest. These weather extremes are more directly related to global warming. Other forms of severe weather that are at least in part greatly enhanced by a warming climate include severe droughts and a rise in extreme summer and winter precipitation events. The effect of climate change on tropical storms, hurricanes, and tornadoes is an active area of research, but the observed data is insufficient to reach definitive conclusions. (20) But it can be said that there is evidence enough to suspect that both the frequency and the intensity of extreme weather events generally are influenced by a warming climate. Even slight increases in the oceans' temperatures can be linked to the intensification of the strength of tropical events. There is some agreement that this is already happening in the Atlantic. (22, 23)

There is absolutely no disputing that the effects of global warming on temperature, precipitation levels, and soil moisture are turning more and more of our forests into kindling during wildfire seasons. Wildfires in the western United States have been increasing in frequency, intensity, and duration since about the mid-1980s. They are occurring nearly four times more often and burning more than six times the land area than they were 30 or so years ago. Wildfires are a natural phenomenon, of course. Cyclical weather occurrences, such as El Niño events, for example, affect the levels of precipitation and moisture, thus producing natural variability in wildfires from year to year. But as the world warms, wildfires occur more frequently and with greater intensity. As conditions grow drier and hotter, forested areas face greater threats from wildfires. Wildfire seasons in the United States are already lengthening, and they are projected to continue to do so as a direct result of global warming. The number of fires and the number of acres consumed is rising steadily. This means more crippling property damages from wildfires that will require billions of dollars more for fire suppression and disaster recovery. (20) As noted in Chapter 1, 2012 saw 9.3 million acres destroyed by wildfires in the United States. Indeed, the average number of acres consumed by wildfires has risen significantly from 1986 to 2012. In 2013, the number of acres consumed was smaller but 2014 saw it rising again. The trend of an increasing number of wildfires and the expanding number of acres involved continues. And the problem is only growing worse, according to most sources.

Climate scientists warn us that we can expect a decline in water quality as a result of our changing climate. This will affect people and ecosystems adversely. Water temperature in the earth's streams, lakes, and reservoirs increases as air temperature rises. This in turn lowers the levels of dissolved oxygen in water and places stress on fish, insects, crustaceans, and other aquatic animals that rely on oxygen. More intense precipitation in some areas will increase runoff that washes more pollution into our waterways. The pollution loads in streams and rivers are naturally carried to larger bodies of water downstream such as lakes, coastal oceans, etc. Among the impacts on human life from this runoff is often the blooming of harmful algae and bacteria. (20) But the greatest concern for humanity may be associated with a basic need we often take for granted.

Perhaps the starkest effect of the anticipated rise in sea level world-wide is its implications for drinking water. The expansion of the ocean as it warms, and the increased melt from ice sheets, ice caps, and glaciers combine to create multiple problems. Along with alarming threats to coastal communities, the constructed infrastructure, economies, and ecosystems already discussed, the rise in sea level has implications for available freshwater. As rising sea levels drive salt-water into freshwater aquifers, the supply of fresh water for drinking is impacted. To be useful for drinking or irrigating, water from our aquifers would need to be treated, usually by energy-intensive and expensive processes. Given the wide range of human activities that depend directly or indirectly on water, future climate-driven changes in water resources can be expected to affect many aspects of our lives. (20)

Whatever its ultimate implications for life on the planet, and many of these will be discussed in detail in the risk and vulnerability assessment discussion of Chapter 4, there is no reason to doubt the science that tells us the climate is changing. It is warming, and this is a problem. The reliable, peer-reviewed climatological research upon which the accepted expert knowledge is based, the evidence, is abundant and robust. Our discussion has touched on some of the major things the science is telling us. The overwhelming majority of scientists who study climate change agree that the earth is warming and the climate is changing. They also agree that human activity is responsible for changing the climate. Irrefutable evidence from around the world—including extreme weather events, record temperatures, retreating glaciers, and rising sea levels—all point to the fact that global warming is happening now and at rates much faster than previously thought. Yet there are doubters and deniers of the science and the things it is telling us. There is no shortage of popular claims contrary to the scientific consensus about climate change.

Among the most widely circulated and popular claims that climate-change deniers will make, there are several types of counterclaims that stand out. The first set of counterclaims holds that the earth is not warming, or global warming has stopped, or that the climate is actually getting cooler. Second is the set of arguments that maintain that while the climate may be warming, it is due entirely to nature or natural processes. The cause of warming and climate variability is natural and not related to human activities or influences. It's all the sun or other natural processes, etc. A third set of claims centers on the notion that global warming or climate change may be happening, but it is not a serious matter of concern. Even if human activities have caused some of the climate change, the future warming that will occur will be nowhere near the projections being made by the climate scientists. While there are infinite variations on these themes, there are perhaps four basic questions that science can answer (has answered, in fact) that pretty much put most counterclaims to rest. Is the climate warming? The science says yes. Are human activities responsible for global warming? The science says yes. Can we make reasonable projections about the future? Again, the scientific answer is yes. Finally, are the impacts and the projected impacts of climate change serious enough to warrant our attention and our action? Yes, again.

Perhaps the best way to discuss global warming denial is to briefly examine in just a bit more detail the answers science has to some of the major denial claims. This, it is hoped, will add to our understanding of what the science is actually saying. But just as it is impossible to comprehensively cover every detail of the science in our summary of it, so too we will not find it possible to be comprehensive in our assessment of the deniers' arguments and the answer of science to them. But we can add to our general understanding of the disagreement between science and denial, a false debate where the science is concerned, with a few examples. Science is what it is, and this is subject to the strictest standards of peer review. As we have previously noted and will always acknowledge, there are uncertainties in science. But there is something called *emerging scientific truth* that has met the test. The arguments that deny it are all skillfully manufactured to argue a point of view, but in most cases, they are not supported by any actual and reliable peer-reviewed science.

Climate Change Denial Claims and the Science

While there are many dramatic claims that deniers of global warming make in an effort to refute the scientific consensus about climate change, the four major ones to be examined here will illustrate an important point that applies to most of them. In almost every instance, the claims of the global warming deniers are either perversions of science, selective presentations or misrepresentations of scientific information, or outright absurdities. Stated another way, there is no scientific evidence that meets the standards of peer review that supports the arguments the deniers make. Let us briefly examine each of four major claims that fall into the denier category.

CLAIM ONE: *Climate change is not occurring (i.e., it's a hoax) or the climate is actually cooling. This is perhaps the easiest of the denial claims to address.*

One of the most persistent and popular arguments in recent times is that the climate is not warming or that it has stopped warming. Another variation says that the climate is actually cooling. The truth is, if we selectively cherry pick start and end dates for short-term observations, it is easy to manipulate information to support any point of view you want. For example, if we fit a linear trend line to global temperature from 1998 to 2008, we find no statistically significant warming trend. However, if we fit a trend line from 1999 to 2008, we do find a strong warming trend. Natural climate variability, including effects like El Niño/La Niña cycles, volcanic eruptions, and solar cycles, may account for year-to-year and short-term variations or wobbles in the overall climate trend. This is why we can observe periods of apparent cooling or periods where there appear to be no trends at all among the last 50 years of global warming. Such variations are expected. (21) This is why climate scientists study trends over a longer period of time. It is neither valid nor illuminating to draw conclusions about where the climate is headed based on selectively targeted short-term observations. But this is precisely what deniers of climate change do when they cherry pick the start and end dates for short-term observations. Any assertions made by anybody on the basis of such observations should be treated with skepticism. More importantly, legitimate climate science leads to the conclusion that the climate is warming, and significantly so. (21)

To determine what the climate is doing, warming or cooling, measurements of temperature or some related quantity must be taken over a long enough time to establish a trend. Each source of relevant data does have its limitations, naturally, but they also have their strengths. It is through the consideration and assessment of a number of data sources that a consistent picture can be drawn. Considered together, and accounting for various short-term influences that may bias the results, these multiple data sources provide decisive evidence that the climate is warming. (1) Among the sources of data are the surface thermometer record, the glacier record, the sea level record, sea ice records, ocean heat content, and satellite temperature measurements. All of these are records of temperatures or something associated with temperature variations. So, for example, the surface thermometer record provides strong and accurate evidence of how much the earth has warmed. But acknowledging the possibility for observational error, changes in the way observations have been made over the past 150 years, and other potential recording errors, scientists have developed techniques to identify and correct them. As a result, the surface thermometer record is today regarded as an important and reliable source. But the conclusion that the earth is warming does not rest on the surface thermometer record alone.

Changes that are observed in glaciers are an excellent indicator of local temperature changes. Warming will cause melting and cause a glacier to shrink. A significant reduction in sea ice is another indicator of global warming, as is the rise in sea level associated with the increased volume of water with warming and the effect of melting glaciers and other ice on land. One of the most obvious indicators of global warming is the amount of energy trapped in the atmosphere by greenhouse gases that contributes to the warming of ocean temperatures. (1) Data gathered from these sources are but a small part of the mountain of data that provides the evidence that the climate is warming. All of this evidence is peer-reviewed and verified by multiple scientific groups. There is always the possibility of error in science, but there is no chance that all of the sources of data examined and all the evaluations of that data could be equally wrong about the overall conclusion that the climate is warming. (1)

CLAIM TWO: The global climate is getting warmer, but not because of human activities; This is a natural process, and human activities do not influence it, etc.

Some deniers will concede that the planet is warming but argue that it is a totally natural phenomenon and there is absolutely nothing we can do that has any impact on it. But what does the science really tell us? To determine what is causing today's increased global warming, scientists have examined all the factors that can affect the earth's temperature. There are essentially three general factors we have already discussed that could be responsible for recent and relatively rapid global warming: 1. The sun; 2. Earth's reflectivity; 3. Greenhouse gases. Scientists have assessed in detail six possible causes of the warming we have observed over the past several decades. These potential causes are variations in the earth's orbit, tectonic processes, volcanic eruptions, an increase in solar output, natural internal variability, and an increase in greenhouse gases. Orbital variations and tectonic processes can be eliminated as significant contributors, according to the science, because they are too slow to have had any discernible effect on climate over the time (decades or even a century) that the climate has been warming. (1) But what about the sun? Isn't it possible that the warming of the planet is entirely due to changes in solar output? The answer is a resounding no.

Ultimately, the climate system is powered by the sun. Indeed, all else being equal, if you turn up the sun, you'll warm up the earth. According to the best and most accurate scientific estimates, the sun has accounted for just a small portion of the earth's warming since 1750. (22) An analysis of the most recent solar activity has in fact demonstrated that since about 1985, solar influence has changed in ways that, if anything, should have had the opposite effect. But global temperatures have been rising. Nothing in the measurements over the relevant time period shows the pattern of changes in solar output that would account for or explain the recent pattern or global warming. Simply put, the sun's output has not been increasing. There is no appreciable change in solar output. In other words, the sun is not causing global warming. (22) In fact, based on solar output, the planet should be cooling.

Internal variations, effects like El Niño/La Niña cycles, do, as we have previously noted, account for many of the observed wobbles and short-term variations we have seen in climate over the past century, but they have been conclusively rejected as causes of the pattern of global warming observed over recent decades. (1) Likewise, volcanic eruptions have been rejected as having any connection to the overall pattern of observed global warming. (1) This leaves greenhouse gases and the greenhouse effect as the only remaining scientific explanation to be explored.

Recalling our discussion of the Keeling curve, we have direct measurements of CO_2 concentrations in the atmosphere going back more than 50 years. We also have many reliable indirect measurements (from ice cores) going back hundreds of thousands of years. These measurements confirm that CO_2 concentrations are rising rapidly. Natural amounts (those not related to human activity) of CO_2 have varied from 180 to 300 parts per million. Today's CO_2 levels have reached over 400 parts per million. That's at least 33 percent more than the highest natural levels recorded over the past 800,000 years. Moreover, increasing CO_2 levels have been proven to contribute to periods of higher average temperatures throughout that long record. (23) We also know the additional CO_2 in the atmosphere today comes mainly from coal and oil, because the chemical composition of the carbon contains a unique "fingerprint" that makes it identifiable.

Increasing anthropogenic greenhouse gas concentrations and the observed increase in global average temperatures are connected. The evidence connecting them is so strong that to claim that global warming was caused by something else, you would have to demonstrate why the observed increase in greenhouse gases is not warming the planet as would be normally expected given what we know about the greenhouse effect. As far as the science is concerned, the case is closed: human activity is causing the earth to get warmer, primarily through the burning of fossil fuels. (23) This takes into account the smaller contribution from deforestation. Given the compelling evidence supporting greenhouse gases and the analysis that confirms the lack of any other plausible alternative scientific explanation for why the earth is getting warmer, it is more than reasonable to conclude that most of the increase in global average temperatures since the mid-twentieth century is due to the increased level of anthropogenic greenhouse gas concentrations. (23)

CLAIM THREE: *The climate is changing, but the impact will be minimal and not dramatic.*

CLAIM FOUR: *There is no reliable scientific basis for predicting the future.*

These two claims can be discussed together. Any discussion about the impacts of climate change by definition includes projecting into the future. Discussing climate-change impacts requires projecting climate change for specific regions and specific seasons. This means projecting not only temperatures but other characteristics of climate, such as precipitation, changes in seasonal cycles, variability, and extremes. The fact is, climate science has documented that the changes already experienced have been significant. Likewise, there is agreement (evidence-based consensus) that the future will present us with even greater challenges.

Reviewing what we already know about the climate over the past 50 to 100 years, there is overwhelmingly compelling evidence of some dramatic changes. Global sea level rose 6.7 inches in the last century (8 inches since 1880). *The rate of sea level rise in the past decade has in fact been nearly double the rate of the past century.* (24) All major global surface temperature reconstructions demonstrate that the earth has warmed since 1850. In fact, the 20 warmest years have occurred since 1981, with the ten warmest years coming in the 2000s even as solar output has declined. (24) Oceans have absorbed much of this increased heat, and ocean temperatures have warmed significantly. The documentation of shrinking ice sheets, the rapid decline in both the extent and the thickness of Arctic sea ice, and glacial retreat almost everywhere around the world indicate dramatic shifts in climate that are impossible to ignore. (24) To this can be added the number of record-high temperatures and the decreasing number of record-low temperatures in the United States (and around the globe) and the changing dynamics of extreme weather events (rainfall events, tropical events, wildfires, flooding, etc.). Finally, consider ocean acidification. Since the beginning of the Industrial Revolution, the acidity of ocean surface waters has increased by 30 percent. Most of this increase is the result of the human emission of more CO_2 into the atmosphere. (24)

While it can be difficult to predict the exact timing and precise extent of all specific future climate-change impacts, there is sufficient scientific knowledge about the general and expected impacts of global warming to cause more than a just little bit of concern. With what we have already observed, with measurable trends already established, and with reasonable scientific projections that can be made from them, the future looks daunting indeed. The Union of Concerned Scientists, based on the analysis of the relevant, peer-reviewed science, has presented a summary of the impacts of global warming that articulates very clearly the scientific basis for legitimate concerns about the future of humanity and of the environment. Science is able to tell us what a warmer world will mean. Let us examine some of the concerns that science is already able to alert us to as we anticipate the future. There is evidence that some of the impacts we shall discuss are already beginning to be felt. There is no doubting the impacts we will feel if the climate continues to warm and our carbon emissions are unabated. Assuming we do not significantly alter our practices and take meaningful mitigation and adaptive measures, it is not a question of if we will feel the most severe of these impacts but of when we will feel them.

First, consider the impacts of a warmer climate on human health. Extreme heat events that will persist for several days or weeks can, as we saw in Chapter 1, be killers. The climate-related projections by science that have proven to be the most robust and most reliable are those that relate to rising temperatures and the impact of them. Higher temperatures are also demonstrably related to and influenced most directly by human behavior. (25) As the planet continues to warm, extreme heat events will become more frequent and, with them, the health risks they impose on the human population. But other projected changes related to global warming will impact human health as well.

Changing temperature and changing precipitation patterns and prolonged heat can create chronic drought. This can be linked to significant increases in wildfires that will place both more residents and more constructed infrastructure in great danger. Because a warming atmosphere holds more moisture, some areas may see extremes in rainfall and flooding, and due to rising sea levels, more people will be put in the path of and endure the risks of stronger storm surges and more severe flooding associated with tropical events. Warmer oceans will also spawn more intense tropical events. (25) Other health-related concerns are tied to declining air quality (higher temperatures will produce more smog), new allergy-related diseases (warmer temperatures and rising levels of CO_2 produces new and more powerful allergens), and climate-related changes in disease vectors (the mechanisms that spread diseases will respond to both higher temperatures and increased CO_2 levels). (25)

Global warming will impact our food and water as well. These impacts may hit the poorest people in the poorest countries first, but they will impact all more quickly than we may wish to think. The productivity of crops and livestock will be adversely influenced. Growing seasons will be altered, and crop yields will decline with higher temperatures, drought-related stress in some areas, or excessive precipitation in other areas. Some areas that now rely on rain-fed agriculture will require irrigation. This could mean a future of higher costs and conflict over water. It will also mean the loss of arable land, as prime growing temperatures may shift to higher latitudes, where soil and nutrients may not be suitable for crop production, as lower latitudes become less and less productive. (25)

Global warming will lead to a decline in drinkable water for a variety of reasons. Extreme rainfall events in some areas may cause municipal sewer systems to overflow and bring untreated sewage gushing into drinking water supplies. Loss of mountain snowpack or earlier spring melts caused by higher temperatures may reduce the availability of drinkable water downstream. Sea level rise will lead to saltwater intrusion into groundwater supplies in coastal areas. Shortages of irrigation supplies can be expected as well as disruptions in power supplies as lower lake and river levels may threaten the capacity of hydroelectric plants. (25) As global warming increases, water may become the most precious and the most costly natural resource on the planet. It could also become the source for new international conflicts and threats to the peace and security of nations around the globe.

The costs of global warming are already staggering and will grow as the climate continues to warm. Sea level rise, floods, droughts, wildfires, and extreme storms will damage more property and require more expensive infrastructure repairs. The impact on homes, roads, bridges, airport runways, railroad tracks, dams, levees, seawalls, and power lines will increase and become prohibitively expensive over time. Disruptions in daily life can mean lost productivity. The warming climate will lead to more lost work and school days and will impose more dramatic and costly harm on trade, transportation, agriculture, fisheries, energy production, and tourism. Climate-related health risks will multiply as well. All in all, a warming climate will complicate, reduce the productivity of, and handicap the human population as it goes about all of its daily business. (25) To even begin calculating the costs of these "inconveniences" in dollars and cents would stagger the mind. Global warming is likely to increase the number of people who will have to leave their homes because of drought, flooding, and climate-related disasters of all kinds. *Climate refugees* can mean mass migrations and new security threats. Mass migrations of people imply, inevitably, the sort of social disruptions that may lead to other serious problems. This would include perhaps civil unrest and the need for military interventions.

Even if our human societies take reasonable actions to cope with the impacts of a warming climate, including the reducing of carbon emissions, the costs of coping with the changes already in motion will be greater than some may imagine. Farmers will need to spend more time and money on irrigation. They will also have to worry about cooling more vulnerable livestock and managing more and more numerous pests. Governments and communities will have to get serious about building houses and infrastructure that are more energy efficient. They will need to build into their emergency management systems early warning systems for heat waves and natural disasters. They will have to improve response capacities to cope with more extreme climate and weather-related events. They will have to expend more resources and effort to make their communities more resilient and sustainable in the face of changing infrastructural and population vulnerabilities. More public funds will need to be dedicated to building seawalls, containing sewer overflows, and improving bridges, subways, roads, and other critical components of the transportation system. Disaster recovery, that is, rebuilding after disasters strike, will prove to be more and more costly. It will also prove to be incredibly expensive and politically challenging to rebuild in ways that may reduce future damages from future and recurring events.

Whether or not one is disposed to take seriously the scientific discussion of projected global warming impacts, those we have already experienced are decidedly influencing and altering the dynamics and nature of climate-related events. The frequency, intensity, and severity of natural storms are changing before our eyes. The number of record heat waves, the growing scope and intensity of wildfires, the frequency and intensity of tropical events, and the damages to our communities imposed by these recurring natural events are quite noticeably impacting our daily lives. Resources for responding to these events are struggling to keep pace with them, and the costs associated with recovery from them are increasing rapidly and approaching unsustainable levels. The implications of a warming climate for the future are increasingly daunting.

What, in the final analysis, should be said about those who deny either the seriousness of global warming as a climate challenge or the need to act in relation to what science has told us about it? *Denial* is a loaded term, of course. Many who disagree most loudly with the mainstream view or scientific consensus about global warming, those who question if it is happening and/or caused by human activity, prefer to call themselves climate *skeptics*. But in most cases, that label is inappropriate. A skeptic is a seeker of truth. A skeptic does not deny science, but when scientifically valid, may legitimately question a conclusion or demand more data, even if it runs against the grain of or is reluctant to accept a prevailing consensus. But most of what is written or said about global warming by those who do not agree with the scientific consensus is more the case of the denial of a truth one doesn't like. It would not be fair to paint with the same brush everyone who disagrees about climate change, but this characterization of climate denial certainly applies to a majority of the argumentation shouted and the ink spilled in opposition to the science and what it is telling us.

The science and the consensus surrounding global warming are resoundingly accepted by every academy of science in the world and by 97 percent of climate scientists. That it isn't accepted by a similar ratio of the general American public is due not to science, but as we will discover in Chapter 3, at least in part to a concerted denial movement and a political effort that promotes it. One can always find a small number of people who deny reality or believe in silly conspiracies. There are people who believe that Elvis is still alive, that aliens from other worlds are among us, that 9/11 was a government conspiracy, that President Obama was born in Kenya, and that Americans never landed on the moon. It is in this same vein that there are some people who deny the scientific consensus surrounding global warming. These people are very small in number. But there are a whole lot more people, very sane and reasonable people, who have been confused and distracted by the politics of the climate change denial movement and who have not been aware that a scientific consensus has existed for a considerable time now. That only deepens the crisis and delays any meaningful effort to manage it.

Global warming is, with respect to the reliable scientific foundation that identifies its already felt impacts and its reasonably projected threats, a long-term threat to both the sustainability and the resiliency of our human communities. It is a threat to our environment. It is also perhaps the single most important issue of the 21st century because of the threats resident in it to bring untold and heretofore unthinkable disasters to our door steps. But is global climate change or global warming an immediate crisis? Is it a crisis right now that requires our urgent attention? The title of this book of course gives away my answer to these questions.

The Climate Crisis

How do you define the word *crisis*? A typical business dictionary definition defines a crisis as a "critical event or point of decision which, if not handled in an appropriate and timely manner, or if not handled at all, may turn into a disaster or catastrophe." (26) Another popular definition, a synthesis of a number of definitions actually, suggests that a crisis "threatens the viability" of an entity or an organization and is of such importance that "decisions must be made swiftly." (27) For purposes of this discussion, and in perfect correlation with my preferred emergency management logic, it is the first of these definitions that we shall begin with as we examine the question of whether the challenges presented by climate change and documented by science have already presented us with a crisis that must be addressed.

To begin with, thinking as I do from an emergency management perspective, neither climate change nor natural disasters are unexpected or unpredictable. Likewise, their causes and solutions are rarely ambiguous. Disasters—natural and all others, for that matter—are quite predictable. They are the predictable result of the relationships between the earth's physical or natural systems, the demographic characteristics of the human community, the built or constructed environment, and the technologies we develop and the decisions we make as we employ them. Floods, tropical storms, excessive heat, and winter storms are all natural hazards that threaten human communities on a recurring basis. But they are not really disasters. The disasters, we might say, are not natural. Disasters are, for the most part, born of human decisions that create the risk for excessive damages and losses that were preventable. It is in this sense that famed geographer and the father of flood plain mitigation, Gilbert White, famously said, "floods are 'acts of God,' but flood losses are largely acts of men." Human action or inaction is the major contributing factor to disaster losses. What we build, where we decide to build it, how we decide to build it, how we live, how we interface with nature, how we pursue our economic development, how we produce and use energy, and just about everything else we do may enhance or detract from the sustainability of our communities and of our world. How we do all of these things may also recklessly introduce or intelligently manage knowable risks, and this, in turn, shapes the nature and the costs of any disaster that may befall us.

With respect to nature, human action or inaction may contribute to the undermining of natural systems that are necessary to sustain humanity and human communities. Our human communities are a subsystem of and dependent on a larger but finite system, the biosphere. Over time, the capacity of the biosphere to provide essential ecosystems (clean air, water, arable land, oil reserves, etc.) will erode or decrease. The growing demands of a rapidly expanding human population and the impact of human activities combine to place greater stress on ecosystems. This intensifies their erosion and contributes to the accumulation of hazards, the incidence of disaster occurrences, and the costs of disaster-related damages.

Many students of population trends have concluded that the current combination of population and consumption is exceeding the capacity of global ecosystems. The natural capacities to continue providing the resources and the services needed for human well-being cannot keep pace with population growth. Many suggest that we have already exceeded the planet's capacity, and as the rate of population growth continues to accelerate, they believe that radical changes to population size and/or consumption levels are needed to achieve a sustainable scale. In other words, population growth is creating serious potential for various disasters and damages. The point here is these threats to our ecosystems are multiplied when one calculates into the mix the accelerating effect of global climate change on changes already being forced by population and growing consumption rates. Even without climate change, our ecosystems are seriously threatened. It is difficult to imagine that we would neither recognize these mounting threats nor take steps to refrain from practices that make them worse, but that often seems to be the state of human affairs.

It seems not unreasonable to suggest that growing per capita consumption rates of an expanding population combined with development strategies and energy requirements that stress the natural environment would contribute to the creation of predictable and severe hazard potentials that place the future health of our human communities and of the natural environment itself at greater risk. Among the impacts of unsustainable population growth, we can identify, if we accept what climate science is telling us, the phenomenon of global warming. Recalling our discussion of the greenhouse effect (the blanketing effect of heat trapping gases on the earth), the Keeling curve (the documentation of the accelerating increase of CO_2 in the atmosphere), and the impact of rising CO_2 rates on climate, it is interesting to see the correlation between population growth and total carbon emissions. As the world's greenhouse gas emissions continue to rise, we know that this makes higher levels of temperature more likely. Likewise, as the human population continues to expand, we know this makes higher levels of greenhouse gas emissions likely. Indeed if, over time, we were to lay the growth of population on top of the Keeling curve and its measure of increasing carbon emissions, we would see a nearly one to one correlation. Population growth has and will continue to contribute to human induced global warming.

My emergency management perspective, as a general proposition, says that *hazard mitigation*, by which is meant the ensuring of resilience in the face of hazard risks and the promotion of economic, political, social, and environmental sustainability, requires a full awareness of hazard risks and vulnerabilities and a plan to reduce them. The prevention of disaster damages, or at least the reduction of their costs to humanity and the environment, is a critical characteristic of a livable and sustainable community. Structural adjustments to create disaster resilience, adaptations to reduce the damage, may enable communities to withstand disaster impacts and recover from them more quickly, but they are not sufficient in and of themselves to promote human and environmental sustainability. It is also necessary to decide to refrain from or avoid activities that constitute direct threats to sustainability. It is necessary to address the causes of the disasters that may befall us. This means an effort to promote human development and living strategies that preserve the socioecological system. This requires that human beings take responsibility for disasters. Identifying hazard risks and the potential for disaster resident in them is step one in taking responsibility.

With respect to our present climate change, global warming, science has alerted us to a number of hazard risks and disaster potentials. The glaciers are in historic retreat, ice caps are melting more rapidly, sea levels are rising, heat is intensifying, seasonal precipitation amounts are changing radically in some cases, storms are intensifying, wildfires are expanding and more severe, and the costs all of these changes and others are imposing on our communities is escalating. The science is presenting more conclusive evidence with the passage of time that the course we are on is unsustainable. Even without a perfect knowledge of what the future will bring, surely these observable phenomena and the solid science behind global warming studies and the impacts of a changing climate suggest a crisis is already at hand. It is not, when one considers the totality of what is actually known and the accumulation of our own recent experiences, unreasonable to suggest that a critical event or point of decision has already arrived. It is compellingly clear from the scientific point of view that this point of decision, if not handled in an appropriate and timely manner, or if not handled at all, may turn into a global disaster or human catastrophe.

Whatever one chooses to believe about climate change and its causes, the evidence is making it more clearly understood by the day that harms, many of them irreversible, are being done by a warming climate. The warming to date may be only a fraction of the heating that is in store for us in the future. Even if carbon emissions fell to zero in a literal heartbeat, the atmosphere would continue to get hotter from the excess heat already absorbed by the oceans due to human activities up to this moment. The global warming we are presently experiencing is a product of our greenhouse gas emissions of the 1970s and 1980s. We have not begun to feel the worst of the heat coming from our emissions over the past 30 years. If the world overheats for a few more decades, and it is likely that it will, over 30 percent of the world will be in drought at any given time. A few more decades of warmth will mean in all likelihood that 50 percent of the land where crops presently grow will be unsuitable for agriculture.

Climate change is not a new invention. For over 150 years, scientists have refined their methods of study and improved our knowledge of it. The science is sound, but that does not mean that it is universally embraced. Science is never universally embraced, and it most of the time will meet with fierce resistance when its discoveries and recommendations are contrary to either our preferences or vested interests. We have seen this play out time and time again. Science produced, for example, conclusive landmark studies on the dangers of DDT, tobacco, acid rain, and ozone depletion. In each of these cases, as is presently the case with global warming, economic and political interests sought to delay recognition of a problem and action to address it. This resistance was aided, as it usually is, by the general public inclination to reject any suggestion that we must adjust our economic and material pursuits in a manner that requires us to defer profits, pleasures, or other benefits to manage long-term threats. But sometimes a crisis, an emergency we have the ability to see coming or have already begun to feel, can inspire action.

From an emergency management perspective, one may conclude that global warming is already happening, its effects are having a negative impact on the communities in which we live and work, and we need to respond. Whatever the broader causes, the scientific consensus, the disagreement with that consensus, and the legitimate scientific uncertainties about the future, enough is reliably known and enough is already being experienced to demonstrate the practical need to factor the scientific realities of a warming climate into consideration as a variable in our economic and community development policies. These must include, as a necessity, the process of preparing for, responding to, mitigating the effects of, and adapting to changing vulnerabilities and threats with respect to natural disasters. In essence, because it is already underway and its effects are already being felt, the time to respond is indeed at hand. We know it is happening, we know its portent for the future, and we absolutely know that waiting for the worst of its impacts to hit before we act is not an option. By that time, by the time the worst happens, it will be too late to respond effectively if at all. That is, from a basic emergency management perspective, an unacceptable alternative for me. We are indeed talking about a critical event or point of decision which, if not handled in an appropriate and timely manner, or if not handled at all, may turn into a disaster or catastrophe. Now is the time to act. But are we ready to act? Are we going to be able to act?

When science tells us things we do not want to hear, we often either ignore the threats it identifies or attack the science itself. This has been the case many times in our history. We sometimes take refuge in opinions that are contrary to the facts. Opinions are not constrained by data or fact. They are emotionally satisfying, and our resort to them is comforting, though not always wise. We may comfort ourselves for a while with our opinions. This is a somewhat natural mental adaptation that avoids unpleasant realities, but we cannot escape unpleasant realities in the end. It is inevitable, of course, that the things science tells us that we do not want to hear have a tendency to come to pass. When they do come to pass, we readjust and refocus. We generally respond with logic and even intelligence when confronted with what cannot, however unpleasant, be avoided. Until that time, we tend to live in an interim that is defined by a false debate.

In the interim, between our first emotional denial of science that tells us things we do not want to hear and the ultimate realization of what the science portends, a war is waged. It is a war waged both within and amongst ourselves. It is not so much a battle between two valid and equally legitimate sides or points of view as it is a pointless battle between thinking and unthinking. It is a debate about engaging or avoiding reality. For a considerable time, we delude ourselves into believing that the two sides or viewpoints locked in conflict are indicative of a debate unresolved. But the contest between science and opinion is not a debate about matters unresolved. It is usually about reality and the denial or avoidance of reality. What we want, what we feel, what we wish to believe, frequently prevails over what we know. That may be the Achilles heel of humanity. Our preference for opinions, wants, and desires, over what we know postpones or delays action that may prevent a disaster. In fact, it often guarantees the taking of the very steps that will create the disaster in the first place.

In the case of global warming, as a thing already happening with already damaging impacts, the last place we need to be is in the interim waging a war between thinking and unthinking, between science and opinion. If this war runs the typical course of delaying or postponing action that may prevent a crisis, it will be much too late. What is already in motion warns us that a crisis is already underway— "Earth, we have a problem." If we wait for the worst to come to pass, it will already have eliminated the possibility for reasonable action. That is the Catch-22 of global warming and the nature of our climate challenge. It's happening before you know it, and it is too late to stop when you become aware of it. If we spend too much of our time now in the interim of the false debate, that time between when science tells us what we do not want to hear and when we are prepared to act, when what is predicted can no longer be denied, it will be much too late to act. But this is very much where we seem to be in the United States, and, to some extent, globally, when it comes to the politics surrounding climate change. We are stuck in the interim when we should be acting with intelligent foresight and a sense of urgency.

Climate change (a.k.a. global warming) is not sneaking up to us unawares. It is not, as we have seen in this chapter, something new or unexplainable. The science is settled, for the most part, and the scientific consensus is that a critical time of decision is at hand. In fact, the time to begin acting is actually long past due. Individually and collectively, humanity has not acted with a sense of urgency. Despite the fact that growing numbers of people are aware of the climate crisis and the need to act in relation to it, serious action often seems not to be an option. The sad reality is that governments the world over have been slow to act in response to climate change. We must, before turning our attention to assessing and responding to risks and vulnerabilities associated with a warming climate, understand why that is the case. The United States certainly presents a fascinating case study of inaction and even political absurdity in the face of what can only be called a global climate crisis. It is to a discussion of the politics of climate change that we now direct our attention. To truly understand and define the crisis, to even begin to do what must be done to manage it, we must factor into the equation as critical variables the willingness or the unwillingness and the ability or inability of governments to act in relation to it and the ability of people to understand it and demand such action.

References

1. Dressler A., and Parson, E.A. (2010). *The Science and the Politics of Global Climate Change*. Cambridge, UK: Cambridge University Press.

2. Christianson, G. (2000). *Greenhouse: The 200 Year Story of Global Warming*. Penguin Books.

3. Stone, B. (2012). *The City and the Coming Climate*. Cambridge, UK: Cambridge University Press.

4. Arrhenius, S. (1896). "On the Influence of Carbonic Acid in the Air upon the Temperature of the Ground." *Philosophical Magazine and Journal of Science* 41, 237–276

5. Lallanilla, M. "What is the Keeling Curve?" http://www.livescience.com/29271-what-is-the-keeling-curve-carbon-dioxide.html (accessed June 12, 2013).

6. Cowie, J. (2013). *Climate Change: Biological and Human Aspects.* New York, Cambridge University Press.

7. National Research Council (2011). *America's Climate Choices.* Washington, D.C., The National Academies Press.

8. Luthi, D., LeFloch, M., Bereiter, B., Blunier, T. Barnola, J.M., Siegenthaler, U., Raynaud, D., Jouzel, J., Fischer, H., Kawamura, K., and Stocker, T.F. (2008). "High Resolution Carbon Dioxide Concentration Records 650,00–800,000 Years Before the Present." *Nature* 453 (7193), 379–382.

9. IPCC (2007). Climate Change 2007 Working Group 1, Summary for Policymakers.

10. Dlugokencky, E.J., Bruhwiler, L., White, J.W.C., Emmons, L.K., Novelli, P.C., Montzka, S.A., Masarie, K.A., Lang, P.M., Crotwell, A.M., Miller, J.B., and Gatti, L.V. (2009). *Observational Constraints on Recent Increases in the Atmospheric CH_4 Burden.* Geophysical Research Letters 36L:L18803.

11. Lean, J.L., and Woods, T.N. (2010). "Solar Total and Spectral Irradiance: Measurements and Models." In *Heliophysics: Evolving and Solar Physics and the Climates of Earth and Space.* Eds. C.J. Schrijver and G. Siscoe, Cambridge. UK: Cambridge University Press.

12. U.S. Global Climate Change Program. http://globalchange.gov/ (accessed June 10, 2014).

13. Peterson, T.C., et al. (2009). *State of the Climate in 2008, Special Supplement to the Bulletin of the American Meteorological Society.* v. 90, no. 8. August 2009, S17–S18.

14. Allison, et al. (2009). *The Copenhagen Diagnosis: Updating the World on the Latest Climate Science.* UNSW Climate Change Research Center. Sydney, Australia, 11.

15. Kwok, R., and Rothrock, D.A. (2009). "Decline in Arctic sea ice thickness from submarine and ICESAT records: 1958–2008." Geophysical Research Letters, v. 36, paper no. L15501.

16. National Snow and Ice Data Center World Glacier Monitoring Service. http://nsidc.org/sotc/sea_ice.html (accessed July 9, 2013).

17. http://lwf.ncdc.noaa.gov/extremes/cei.html (accessed July 9, 2013).

18. http://www.pmel.noaa.gov/CO2/story/Ocean+Acidification (accessed July 9, 2013).

19. Sabine, C.L., et al. (2004). "The Oceanic Sink for Anthropogenic CO_2." *Science* 305, 367–371.

20. Union of Concerned Scientists. http://www.ucsusa.org/global_warming/science_and_impacts/impacts/ (accessed July 9, 2013).

21. Easterling, D.R., and Wehner, M.F. (2009). "Is the Climate Warming or Cooling?" *Geophysical Research Letters* Vol. 36, No. 8.

22. Lockwood, M., and Frohlich, C. (2007). "Recent Oppositely Directed Trends in Solar Climate Forcings and the Global Mean Surface Air Temperature." *Proceedings of the Royal Society* Vo. 463 No. 2086, 2447–2460.

23. Boden, T.A., Marland, G., and Andres, R.J. (2012). *Global, Regional, and National Fossil-Fuel CO_2 Emissions.* Carbon Dioxide Information Analysis Center, Oak Ridge National Laboratory, U.S. Department of Energy, Oak Ridge, Tenn., U.S.A. doi:10.3334/CDIAC/00001_V2012.

24. NASA. The Evidence for Rapid Climate Change Is Compelling. http://climate.nasa.gov/evidence (accessed July 9, 2013).

25. Union of Concerned Scientists. "Impacts of Global Warming." http://www.climatehotmap.org/global-warming-effects/ (accessed July 9, 2013).

26. Business Dictionary.com. http://www.businessdictionary.com/definition/crisis.htm (accessed July 10, 2013).

27. Pearson, C., and Clair, J. (1998). "Reframing Crisis Management." *Academy of Management Review* 23 (1), 59–76.

CHAPTER 3

The False Debate: A Confused Public

Introduction

If the science on global warming is settled, what is to be made of the considerable public and political disagreement in the United States over the subject? Why, if the preponderance of evidence and the consensus of climate scientists is to be believed, is there such intense disagreement in political circles and amongst the public in general about the need to respond to the warnings ("Earth, we have a problem") that the science has so clearly demonstrated? Understanding the answers to these questions is a critical necessity if we are to make any progress in addressing the challenges presented by climate change. The climate crisis cannot be managed if it is not perceived to be a crisis by political actors and the general public. Managing the crisis cannot happen without an engaged public and without a supportive public policy environment.

As we have discussed in Chapter 2, knowledge about climate change, like all scientific knowledge, is subject to some uncertainty. Projections about the future of global warming are also subject to uncertainty. But as we have attempted to demonstrate in our discussion of the science, uncertainty does not imply a lack of scientific evidence that is reliable with respect to the identification of trends and probabilities. Also, we must be careful to understand that uncertainty cuts two ways. The ultimate climate changes and their impacts may be smaller or larger than any projections drawn from the estimates that may be based on the best available and most reliable scientific evidence. Indeed, it is important to note that most scientific projections about the climate tend to be on the conservative side. Scientists are wary of endorsing dramatic scenarios and tend to be cautious with their projections. This means the chances are greater that they underestimate climate-change impacts to some degree. But even the most conservative or cautious estimates of future impacts of a warming climate do not justify a course of doing nothing or waiting to see what will happen. Just as we do not wait for an illness to become life threatening before treating it, the science makes very clear the foolishness of waiting to see how bad the impacts of global warming will be before responding to them. But as a policy matter and/or a political matter, the science appears to be limited in its ability to influence the necessary consensus in the public and political arenas to create the best possible relationship between climate science and public policy.

It is a given that, when policy issues have high stakes, political debate in the United States will be contentious. Because the potential impacts of global warming are so threatening, because the fossil-fuel consumption that contributes to it is so important to the world economy, and because the costs of addressing it and/or ignoring it are so very great, strongly felt and opposing views will be expressed and argued in our public discourse. In fact, the number and the intensity of contradictory claims advanced about climate change are extreme, as we see in the following list of quotations. Whether these claims and counter-claims are based on the best and most reliable scientific evidence is not generally considered by the average citizen. Absent any significant background or familiarity with the scientific basis for evaluation, the average citizen is frequently left with the mistaken impression that there are two sides to the climate debate that are more or less equal and that the science itself is in dispute.

Sample of Conflicting "Views" in the Public Discourse

"The danger posed by war to all of humanity—and to our planet—is at least matched by the climate crisis and global warming. I believe that the world has reached a critical stage in its efforts to exercise responsible environmental stewardship."
—UN Secretary General Ban Ki-moon, Remarks to Bali Conference on Climate Change, December 12, 2007

"Ambiguous scientific statements about climate change are hyped by those with a vested interest in alarm, thus raising the political stakes for policy makers who provide funds for more

science research to feed more alarm to increase the political stakes."
—Professor Richard Lindzen of the Massachusetts Institute of
Technology, op-ed *Wall Street Journal*, April 12, 2006

"I want to testify today about what I believe is a planetary emergency—a crisis that threatens the survival of our civilization and the habitability of the Earth."
—Al Gore (testifying on impact of global
warming before U.S. congress)

"Environmental organizations are fermenting false fears in order to promote agendas and raise money."
—Michael Crichton

"Global warming is too serious for the world any longer to ignore its danger or split into opposing factions on it."
—Tony Blair

"With all of the hysteria, all the fear, all the phony science, could it be that man-made global warming is the greatest hoax ever perpetrated on the American people? It sounds like it."
—Senator James M. Inhofe, remarks on the
Senate floor, July 28, 2003

Global warming in American public and political discourse has not been perceived as a settled matter. It has been most inaccurately perceived as a controversy, scientific as well as political, in the eyes of many casual observers. Debates over whether global warming is occurring, disagreements about its causes and effects, questions about the legitimacy of the science itself, and disagreement over policies and actions that might be necessary to respond to it have led many Americans to believe that the science itself is not after all decided, that there is no overwhelming scientific consensus about the need to act. This has created room to promote doubt about the science in the public mind. In the summer of 2013, President Obama expressed some impatience with those who continue to promote doubt about the science:

So the question is not whether we need to act. The overwhelming judgment of science—of chemistry and physics and millions of measurements—has put all that to rest. Ninety-seven percent of scientists, including, by the way, some who originally disputed the data, have now put that to rest. They've acknowledged the planet is warming and human activity is contributing to it. . . . Nobody has a monopoly on what is a very hard problem, but I don't have much patience for anyone who denies that this challenge is real. We don't have time for a meeting of the Flat Earth Society. Sticking your head in the sand might make you feel safer, but it's not going to protect you from the coming storm. And ultimately, we will be judged as a people, and as a society, and as a country on where we go from here. (Remarks by the president on Climate Change, Georgetown University, June 25, 2013, http://www.whitehouse.gov/the-press -office/2013/06/25/remarks-president-climate-change)

The president's lack of patience notwithstanding, he and other policy makers who might be inclined to acknowledge the scientific reality of the climate crisis and to address it on a policy level have found the political terrain to be very daunting. Some would say this has discouraged action, or at the very least, that the political and economic forces opposed to action have been too strong to overcome. Indeed, the president himself has been inconsistent in addressing the climate challenge he clearly sees as critical in the "overwhelming judgment of science." In the face of an ideologically divided Congress, more intensely divided and dysfunctional than ever, with public opinion on global warming falling into warring partisan camps, and in the face of intense lobbying efforts by well-funded industrial and economic interests who fund the political process (i.e., campaign donations) and who advertise with independent expenditures to influence it, progress is in short supply. Political figures, including President Obama, have all too frequently evaded action on the climate crisis. At least it may be said that they postponed action in the interest of their political survival.

The doubts, disagreements, and confusion of the American public and political decision makers about global warming are really most peculiar in the context of the past 150 years of scientific research and study of climate change. It is disturbingly accurate to say that our public policy and our public discourse have both lagged very far behind the science. Indeed, as the science has become more settled, it seems that the policy and political components of the climate discussion have become more unsettled and contentious. This has been, we shall see, a matter of design, as many participants in the public dialogue have sought to create doubt, sow seeds of confusion, misdirect the public, and profit by the delay brought about by disagreement.

In American politics, when there is a debate, genuine or manufactured, there is a no-holds-barred contest to influence policy outcomes. There is nothing wrong with such a contest in a free society, of course. The political contest often includes aggressive and effective dissemination of information that is less than accurate or truthful. But that's politics. That's why we want to hear both sides of a debate. That's why, in the political realm, we often assume that the competing or disagreeing sides suggest a contest between equally flawed partisans and the truth, whatever it may be, is to be found somewhere between the two sides. This is often the case in politics. But the political debate about global warming is pretty much a false debate where the science is concerned. The science is not very much in dispute. The political debate is a manufactured disagreement to some extent promoted by efforts to create doubt or confusion about the science. Some of this effort is expended to suggest that the science itself is politically motivated, and as such, suspect. The illusion of a contest over scientific matters unresolved perpetuates the notion that neither "side" in the climate change debate can be entirely believed. This reduces science and reliably peer-reviewed findings to the level of other mere biased partisans in the discussion and perpetuates disunity and inaction in the face of the overwhelming scientific evidence and its warnings. In many ways, it is perhaps most accurate to suggest that the divisions that appear in our current politics over the issue of climate change are entirely the result of things having absolutely nothing to do with science.

From Climate Science to Climate Politics

In 1965, the President's Science Advisory Committee produced a summary of the potential impacts of carbon dioxide on climate. This summary reached the conclusion that increasing levels of CO_2 in our atmosphere (25 percent more was being projected by year 2000) would lead to modification of the "heat balance of the atmosphere" that could contribute to significant changes in climate. (1) This report was referenced in a Special Message to Congress delivered by President Lyndon Johnson in February of that year. In reference to the scientific conclusions, he stated, "This generation has altered the composition of the atmosphere on a global scale through radioactive materials and a steady increase in carbon dioxide from the burning of fossil fuels." (2) This statement was part of a broader statement of environmental initiatives designed to address what the president referred to as the "darker side" of modern technology and to deal with its "uncontrolled waste products" that were "menacing the world we live in." (2) Buried within a larger package of environmental policy initiatives, the mention of the impact of CO_2 produced by fossil fuels on the atmosphere was perhaps little noted. But it was not disputed either. Through most of the next decade, it received little attention in the policy arena. Given the importance of and national preoccupation with issues such as the Vietnam War and the Watergate scandal, this is no surprise. But this would change by the end of the 1970s. In 1977, the Department of Energy (DOE) reviewed its research programs related to CO_2, including the possible impacts of changing CO_2 levels on climate. Over the next couple of years, a select group of scientists conducting the DOE review showed that increases in the carbon dioxide concentration of the atmosphere would result in an increase of average surface temperatures. Of greater concern than the general temperature increase associated with rising CO_2 levels was the likelihood that the warming experienced would be greater, perhaps a good deal greater, at the poles. (3) President Carter's science advisor asked the National Academy of Sciences (NAS) to convene a panel to review the findings of the DOE review. This NAS panel included two leading climate scientists, Syukuro Manabe and James E. Hansen, and their state-of-the-art climate models. The conclusions reached by the NAS panel concluded that the climate is indeed sensitive to changing levels of CO_2 (i.e., warming is likely with rising CO_2 levels). There are natural processes that might act as a brake on warming, but the study noted that this was not enough to prevent significant warming. Available evidence suggested that ocean mixing, or the thermal inertia of oceans, the distribution of heat into deeper waters thus slowing their heating, would slow the warming of the oceans and delay the ultimate warming of the atmosphere. But within several decades' time, approximately 50 (perhaps fewer) years according to this NAS study, the effects would be pronounced and have major impacts on the climate. (4) This study led to requests for more analysis and information, especially with respect to the timing of any potential negative impacts. A demand for more study and more scientific precision was logical and desirable. But this is not the only response to the NAS study. It would also lead, over the next decade, to the introduction of economic and political variables into the conversation. This often aided attempts to shade or weaken what the science was able to say with the passage of time and continued research. This served to delay or postpone immediate policy initiatives or significant action.

In 1978, in response to what the science was showing up to that time, the U.S. Congress enacted the National Climate Act. This included significant funding for climate research and authorized the National Academy of Sciences to undertake a comprehensive study of CO_2 and climate. (5) The introduction of climate and the impact of CO_2 on it, the introduction of the science surrounding global warming to the political and policy agendas, was a necessary and inevitable thing. But it is important to note that this development meant that the climate would now, in addition to its being a scientific issue, become a political issue. As a political issue, it would be influenced by much more than science. Economics, political ideology, conflicting interests, and all of the emotive and partisan concerns that animate our political life in the United States would now influence the discussion and the debates to follow as much, more perhaps in some cases, than climate science. Politics in our republic is about competing to influence outcomes and advancing political and economic interests, not about scientific consensus. It is, in the United States at least, also a contest to influence the mood and the opinions of the public. This contest is waged to influence election and policy outcomes in one direction or another. This effort to influence political outcomes has the potential to skew the scientific conversation both ways. Some special interests and political entities may work to suppress or deny the scientific consensus about global warming, and a few others may work to amplify the alarm about it. Well-funded special interests promoting economic causes, fossil-fuel interests, energy policy, and industrial innovations routinely engage in skillful lobbying efforts to shape, delay, promote, or prevent climate policy initiatives. This is basic economic self-interest. But it also changes the nature of the conversation to be had. With the introduction of climate to the public policy agenda, we aren't in Kansas anymore (i.e., it's not just about the science anymore). It is perhaps safe to say that even though climate change has made it onto the political agendas of the 1980s, 1990s, and the first decade of the 21st century, little real progress has been made in the policy debate.

The National Academy of Sciences, pursuant to the 1978 Natural Climate Act, conducted a comprehensive study of CO_2 and climate. This study included, in addition to the science, a comprehensive assessment of all aspects including an economic assessment of CO_2

emissions and the effort to reduce any climate impacts. The final report of this study presented two very different perspectives, one scientific and one economic, with the economic perspective dominating in some respects. The parts of the report written by scientists were consistent with what science had already said. Most importantly, nobody challenged the conclusion that the climate was warming and that this would have serious impacts on the environment. The most likely scenario, based on the analysis of the scientists, was a doubling of CO_2 in the atmosphere by 2065. They added that it would be unwise to exclude the possibility that this would happen earlier, in the first half of the 21st century, and that there would be observable and negative climate impacts. (6) It was true that the climate is naturally variable, said the scientists, but the rapid and forced change being projected due to CO_2 emissions related to increasing fossil-fuel consumption were human (i.e., anthropogenic) causes that had the potential to seriously challenge ecosystems in just a few decades' time and adversely impact human life. This was the primary concern of the scientists. To address the threats that climate change (warming) would cause, things like increased taxes on fossil fuels and other regulatory measures to discourage and ultimately reduce fossil-fuel consumption and carbon emissions were mentioned. The other options included adaptation to a higher-CO_2 world and its higher temperatures. The scientists preferred efforts at reducing carbon emissions. (6) But the economists had a different set of concerns and recommendations.

The economists did not disagree with the scientific facts, but they did disagree with the interpretation of them. Stressing that CO_2 was not the *only* cause of climate change (even though the scientists were saying it was in fact the major cause), the economists thought it would be wrong to commit to the notion that fossil fuels and carbon dioxide were either the main cause of the problem or the area where a solution must be found. There was considerable uncertainty about both the extent and the timing of any of the negative impacts described by the science. The economists therefore thought it premature and unwise to think in terms of action, rejected as unproven the scientific conclusion that any future impacts would be potentially severe, recommended more study, and recommended doing nothing in the near term. (6) The economists concluded that there was no political or economic solution to the problems associated with a warming climate, assuming for the sake of argument that it would present serious problems that merited serious attention, because they were not workable in the end and they would do great economic harm. The greenhouse effect could not be stopped, so the best approach or only choice was essentially to adapt to higher temperatures. They expressed confidence that if, indeed, carbon emissions were to prove to be the primary cause of global warming, and if the impacts of a changing climate were to become severe—and they did not think they would, even though they provided no evidence to substantiate their belief or refute the scientific concerns—the free market would ultimately resolve the matter. (6) The final report from this study emphasized the economists' interpretation and muted the scientific component. This effectively eliminated any felt need to be proactive in addressing or limiting carbon emissions. It also defined the nature of the "debate" that would unfold over the coming decades.

On one "side" of the climate "debate" is climate science. On the other have been a host of influential political and economic actors whom I will label as the "not-science team." This is not to suggest that climate science has had a political agenda or that the "not-science side" has always denied scientific conclusions or been anti-science. It is rather to suggest that climate science versus the not-science team (i.e., the host of political and economic actors motivated by interests other than science) is the place where the basic debate has played itself out, and, one might add, to the advantage of the not-science team, since it is really the only purely political player in this particular contest. Essentially, the debate we have been having in the United States for the most part has not been a scientific debate. The science is what it is at any point in its inquiry. Over time, as demonstrated in Chapter 2, the science has become settled. A scientific consensus has been achieved with respect to global warming, its causes, its already observed impacts, its anticipated or future impacts, and about the need to act. Increasingly, as the science has advanced and the evidence has led to scientific consensus, the "climate debate" has evolved into a contest between the conclusions of science and the devaluing, questioning, doubting, and ultimate denial of science for the explicit purpose of protecting economic and other interests that feel threatened by any actions that might be taken to respond to the scientific warnings. Over time, the "not-science team" adopted the strategy of politicizing the science, denying the science, even demonizing the science in order to win the political debate. As the science became more settled, the work to discredit it accelerated and resulted in what some would call a war against science. The "not-science team" inevitably morphed into an "anti-science" entity in the service to its array of other economic, ideological, and political objectives.

Influential economic and political actors (the not-science team) have consistently argued that the scientific work was incomplete, filled with uncertainties, and too weak to justify taking policy action. That tactic was there at the beginning and has remained throughout. But as the science became more and more settled over time, these economic and political actors escalated their efforts to promote or manufacture not just skepticism or uncertainty about the science and its conclusions but a full-fledged denial of it. The opposition of the fossil-fuel industry and other industries eventually produced well-funded lobbyist groups and public relations campaigns to promote the anti-science argument (escalating from skepticism to denial) in order to slow or prevent potential governmental regulation. That, and not the science, is what their side of the "debate" is really all about. It is about slowing or preventing a policy response to the science that might be damaging to their economic interests.

One of the greatest ironies, of course, is that some of the "anti-science" fueling the denial of climate change was perpetuated by a few fairly renowned scientists. These scientists, award-winning scientists in their respective fields in some cases, were important in conducting research on Cold War weapons systems. They were active weapons systems researchers and staunch supporters of President Reagan's Strategic Defense Initiative in the 1980s. However, none were engaged in climate research, nor were they experts in that field. None were climate scientists. But in the waning days of the Cold War, their life-long commitments to the free enterprise system and the national security of the United States, and in their desire to remain relevant to protecting both, drew them to focus on what they increasingly perceived (ideologically speaking) to be the new great threat to liberty. Environmentalism was, in the view of these prominent and influential defense-oriented scientists, the new great threat to free market capitalism. All forms of governmental regulation were thought to be suspect by this patriotic lot of Cold War scientists. Environmental regulation was, in the aftermath of the Cold War, the greatest threat to liberty in their eyes. The entire climate change or global warming debate, especially with policy implications that would potentially include regulating or taxing carbon emissions, had to be rigged to protect free market capitalism. (6)

Many of the ideologically motivated scientists waging war against climate science began their campaign of strategic action and disinformation through an entity called the George C. Marshall Institute. Sponsored by Exxon and related entities, this institute is an extremely conservative, corporate-funded front group that pursues its political agenda under the guise of promoting reliable or sound science in policy areas where science and technology are major variables. Their scientific perspectives on climate change would, of course, have nothing to do with climate science and everything to do with discrediting it through nonscientific attacks and unscientific conclusions dressed up as science, although they are in direct contradiction to it. (6) Every one of their publications began with the conclusion that climate science is filled with uncertainty, undecided, and provides no basis whatsoever for policy action. (7) But virtually everything they published was in fact contradicted by the real, established, and accepted scientific evidence published in the actual peer-reviewed research that had been done by the actual experts. But then, just as their goal was to discredit the science, their audience was not a scientific one. The objective was to reach the public, distort perceptions, discredit the science, manufacture doubt, and delay action. Little of their "work" on the subject has been actually published and reviewed in scientific publications. Most of it has been for the mass media and for consumption by the uninformed public they seek to manipulate.

A powerful ally to the "not-science team" now morphed into an "anti-science team," though unwittingly so in some cases, has often been the American news media. On the pretense of being fair and balanced in its reporting, the media, by presenting "sides" and rarely any "facts," has helped create over the past three decades the inaccurate public perception that there was a major division among climate scientists regarding the facts or the causes of global warming. Some media outlets did, and still do, little more than serve as a bullhorn for the promotion of doubt about the scientific evidence and the scientific consensus about global warming. (8)

By the late 1980s, the battle lines between climate science and the "not-science team" (soon to evolve into an "anti-science team") were well established, and the unfolding political debate followed a predictable and steady path. In 1988, James E. Hansen, director of the Goddard Institute for Space Studies, announced that the science was clear that anthropogenic global warming had begun. In 1989, the Marshall Institute responded by issuing its first "report" disputing climate science and attacking its findings. The initial strategy was not to deny the fact of global warming. Rather, it was blamed on the sun. (6) As we saw in our discussion in Chapter 2, the science has proven that global warming is not being caused by the sun. There is no doubting the anthropogenic causes and the linkage of global warming to fossil-fuel consumption. But that fact matters little. Usually, the not-science cum anti-science argument has been able to manipulate the legitimate uncertainties in the science, and combining them with enough misinformation or distorted information, create enough disagreement or doubt to prevent policy makers from taking significant action. But there are exceptions.

As the science does become more certain, or where unfolding events may periodically contribute to political pressure to take climate change seriously, there have been spurts of constructive attention by policy makers. But the story for much of the past thirty-five years has been one of increased political disagreement and inaction on the policy front. Scientific advancements in the understanding of the climate crisis have been countered with stiff political resistance that has delayed or retarded progress on the policy front. The politics of climate change has created a political climate of doubt and a legacy of inaction, at least insofar as the American policy agenda is concerned. It is worth recounting, however briefly, the blow-by-blow of the political fight over climate change over the past three decades.

Science, Politics, Doubt, and Inaction

In February of 1979, the World Meteorological Association (WMO) held the first world climate conference. Based on the scientific work completed at that time, this conference was the first major global recognition and discussion of the human contribution to climate change. In fact, this WMO conference would provide the foundation for the United Nations panel that would begin to systematically study the issue of climate change nine years later. Noting that carbon dioxide played a fundamental role in determining the earth's surface temperature, the WMO conservatively concluded in 1979 that "it appears plausible that an increase in the amount of carbon dioxide in the atmosphere can contribute to a gradual warming." (9) Concluding that the details were, at the time, poorly understood, the WMO emphasized the need for further study. By 1988, the scientific study of climate change had advanced sufficiently to encourage global policy makers to begin taking more than a casual interest. The United Nations Climate Program and the World Meteorological Association established the Intergovernmental Panel on Climate Change (IPCC). Its assigned task was to assess available scientific data and the possible impacts of climate change. This was meant to determine if a global response might be necessary to address any major and identifiable risks and vulnerabilities. (10) The IPCC involved hundreds of scientists. They were organized into working groups, each assigned to assess a different aspect of the issue (the atmospheric science of climate change, the impacts of climate change, adaptations to a changing climate, and the possibility of reducing the greenhouse gas emission that contribute to global warming). (11) Since its formation, the IPCC has issued five major assessment reports (1990, 1995, 2001, 2007, and 2014) that are regarded as authoritative summaries of the state of scientific knowledge about climate change. As the IPCC began its work in the late 1980s and early 1990s, governments around the world also placed climate change on their agendas. High profile international conferences called for the reduction of worldwide carbon emissions. The goal was set at a 10 to 20 percent reduction in the rate of carbon emissions. A Framework Convention on Climate Change (FCCC) was signed in 1992 (the first international treaty on climate) and went into effect in 1994. One hundred and ninety nations, including the United States, ratified this agreement. With the goal of stabilization and eventual reduction of greenhouse gas concentrations, the FCCC provided a starting point for negotiations to identify and achieve more specific and binding measures. Nations committed, under the FCCC framework, to voluntarily adopt and report on measures to limit greenhouse gas emissions. Few governments, however, made much progress in reducing emissions or taking concrete action to achieve that goal. By the mid-1990s, the need for a plan and a binding international agreement to reduce carbon emissions seemed both necessary and feasible. Negotiations along these lines led to the 1997 Kyoto Protocol. (11)

The Kyoto Protocol sought to reduce emissions of greenhouse gases to 15 percent below 1990 rates by 2010. Thirty-seven industrialized nations, including the United States, were required to reduce greenhouse emissions. The larger burden fell to developed countries because, as the treaty argued, they had more responsibility for the then current level of pollution. (12) Initially the Clinton Administration had opposed near-term emissions cuts. Instead, the United States proposed only research and voluntary reductions in the early years and a delay on any emission targets until 2008. The United States Senate, thinking the treaty was unfair (i.e., economically harmful) to developed countries and hostile to any emission limits unless developing countries made commitments to emission cuts at the same time, passed a resolution that rejected emission standards for developed countries. (10, 11)

The Kyoto Conference reached an agreement that imposed specific emission targets for each industrialized country over a five-year period from 2008–2013. The treaty included no emission limits for developing countries. Despite this, and in spite of the Senate resolution rejecting emission standards for developed countries, the U.S. delegation signed it. (11) The Kyoto treaty agreement was essentially a Protocol that established emission limits but that allowed flexibility in how nations might choose to achieve these limits. Most of the details with respect to the implementation of the Protocol were left to be settled by subsequent negotiations.

In addition to political disagreements between developed and developing nations on the nature of emissions limits, the Kyoto agreement produced tremendous push back from various corporate and economic interests that might be impacted by serious efforts to limit greenhouse gas emissions. In April of 1998, and very much in reaction to the Kyoto Protocol, global warming skeptics (the not-science team) developed a plan of action. Oil and gas giants like Exxon-Mobil had lobbied extensively against the Kyoto Protocol on the grounds that it would be too expensive to implement, placed too much burden on developed nations, and would have devastating economic impacts. Lee Raymond, chief executive of Exxon-Mobil, was also convinced that the science behind global warming had to be wrong, or at least he perceived it to be such. Thus, Exxon began to fund research groups (e.g., George C. Marshall Institute) to challenge the science. One such group, the Global Climate Science Team, drew up a plan to challenge the science behind climate change claiming that "victory" would be achieved "when average citizens understand or recognize uncertainties in climate science." The recognition of these "uncertainties" must be made to shape and dominate the public's conventional wisdom about climate change. (13) The ultimate goal, of course, was to systematically misrepresent uncertainties to such a degree that they would discredit all of the science behind climate change and delay or prevent any policy responses to it that might be harmful to the fossil-fuel industry's bottom line. (13)

It is fascinating to pause and reflect on the preceding paragraph. Based on the belief of its CEO (not any science) that the science is "wrong," Exxon-Mobil wishes to fund research that will support what it believes. It begins with the conclusion that the science is wrong, funds groups willing to prove what it believes, and these groups think in terms of a "victory" that will create public doubt about what the science has actually been able to demonstrate by making its "uncertainties" (some real but many also exaggerated or alleged) the sole focus of public attention. This is not science! It is not how the scientific method works. This is corporate financing to influence public opinion. It is politics. As a political tactic, it is legitimate and of course is frequently, that is, every single day, employed. But it is not science. Science does not begin with a conclusion based on an opinion by an interested party who funds research to support what it believes. The ensuing debate, a political debate to be sure, will not be a scientific one. But then politics is not science, and science is not politics. The science is what it is at any moment. With respect to climate change, as we saw in Chapter 2, the science has pretty much been settled, and for some time. The politics has been, and—very much by design—remains unsettled.

In April 1998, the Oregon Petition (organized by a small entity called the Oregon Institute of Science and Medicine) began to collect signatures from "scientists" who "believed" that there is no convincing scientific evidence that climate change is man-made. The petition has attracted over 30,000 signatures (it is still gathering them). Very few are climatologists (just over 30 claimed to be active in the field) and many are joke entries of questionable scientific pedigrees (e.g., Major Frank Burns of MASH fame and Geri Halliwell, better known as Ginger Spice). Despite its large number of questionable, joke, and in many cases fraudulent, names and its general lack of legitimate and active climate scientists, this petition is frequently cited by deniers as evidence that there is a lack of consensus in the "scientific community" about climate science and there is a raging scientific debate about it. (14)

As the 1990s came to an end, and with the beginning of the Bush Administration in 2001, the United States became even more reluctant to bear the burden of any costs associated with precautionary policy actions to address the challenges of climate change. Focusing on the "uncertainties" (and as we have stipulated in Chapter 2 and throughout our discussion, there are always some uncertainties) in climate science, the Bush Administration took the extreme and frankly ridiculous position of demanding absolute scientific certainty before accepting any claims that human-caused climate change required any precautionary action or policy response. (15) This position, which essentially echoed the position of industry, especially the fossil-fuel industry, and climate change skeptics and deniers, foreshadowed the inevitable decision of the Bush Administration to withdraw from the Kyoto Protocol. On March 29, 2001, President Bush announced that he would not send the Kyoto Protocol to Congress for ratification. White House spokesperson Ari Fleischer, in making the announcement, said, "The president has been unequivocal. He does not support the Kyoto treaty." (16) Three reasons were cited in support of this decision. First, adherence to the Protocol would impose regulations that would be harmful to the U.S. economy. Secondly, it was unfair that the burden fell upon the United States and other developed countries. It was not fair, in other words, to let developing nations (including China) opt out of mandatory carbon restrictions. Thirdly, the science was said to be "uncertain," and, as such, did not justify moving forward with expensive precautionary steps and policies. (15, 16)

Would ratification of the Kyoto treaty have harmed the U.S. economy? That would depend on the variables to be considered. But the Bush Administration did not consider the considerable costs of doing nothing either. The fact is, making the assumption that doing nothing is cost free or that there would be no economic benefits to implementing the Kyoto Protocol is extremely dubious. The potential for growth in "green jobs" and new industry to propel carbon reduction and clean energy alternatives was never explored or considered. Indeed, instead of seeing the proactive effort to reduce carbon emissions and alternative renewable sources as costs, they could be seen, as the Obama Administration would later argue, as an opportunity to create a new sector of "green jobs" and a means of protecting and enhancing American global competitiveness. (15)

Was it "unfair" to place a greater burden on developed countries such as the United States? Considering total carbon emissions by country, the total emissions of China were at the time close to those of the United States and would continue to grow and surpass the United States with economic development. It might seem that the U.S. contention of "unfairness" is justified. But if, instead, *per-capita emissions* are considered as a variable, the contributions to the carbon problem by the United States and other developed nations are much higher than the developing world. The decision by the UN and the Kyoto Protocol to base agreements on per-capita emissions is justified by the fact that the developed nations have contributed much more to global emissions and the problems associated with them, and they are better situated to begin addressing the problem. And the economic reality is that the developing world needs to continue to develop to the advantage of all nations. (15)

With respect to the Bush Administration's contention that the science was too "uncertain" to justify action, this is more of a political argument than it is a scientific one. The science had, by the beginning of the 21st century, as we discussed in Chapter 2, produced ample evidence to justify policy action. But economic interests fearful of being on the "losing side," should carbon reduction measures be adopted, fought the science tooth and nail. Groups funded by big oil (e.g., The Global Climate Coalition, the American Petroleum Institute, energy producers, automotive companies, industry groups of all sorts) and anti-science skeptics and deniers all supported the efforts to weaken the science and argue that its "uncertainties" advised against any policy action. Some groups, including the George C. Marshall Institute, the Competitive Enterprise Institute, and the Center for the Study of Carbon Dioxide and Global Change, concentrated all of their efforts on discrediting climate science. Exxon-Mobil, for example, funded more than 40 such groups between 1998 and 2005 at a cost of $16 million. These nonstop disinformation efforts—and that is exactly what they were and are—combined with campaign contributions and extensive lobbying efforts have been very effective at delaying any policy action in the United States (15)

The United States was not alone in questioning the Kyoto Protocol. Many of the industrialized nations shared the same "concerns" about the economic impact of the Protocol and the fairness of it, though none questioned or sought to discredit the science with the same intensity as the United States. The Kyoto Protocol went into effect on February 16, 2005. But without the participation of the major per-capita emissions producers, including the United States and Russia, it is essentially a symbolic agreement between the 141 nations who ratified it. (17) Negotiations have continued, but progress remains limited. Yet, and despite continued political division, the conversation may be starting to change. There may be at least a few signs of change, small but hopeful, on the horizon. In March of 2006, Rex Tillerson became Exxon-Mobil's new CEO. On the heels of spending $16 million to discredit all of the science about climate change, Tillerson stated in an interview, "We recognize that climate change is a serious issue. We recognize that greenhouse gas emissions are one of the factors affecting climate change." (18) While a cautious step, this statement was followed in December 2007 by an Exxon-Mobil announcement that it would discontinue contributions to research groups that questioned or sought to deny climate change. By January of 2009, Exxon-Mobil was advocating policies to reduce carbon emissions. CEO Tillerson announced that Exxon-Mobil now supported a carbon tax. He also stated that greenhouse gas emissions were a problem that would likely increase and that managing the risks imposed by greenhouse gas was one of the major challenges of our time. (17, 18)

Exxon-Mobil's retreat from the anti-science or not-science team, encouraging as it is, did not leave much of a void in the efforts to attack or discredit the science. Exxon-Mobile still had and used back doors to slip their dollars through to support climate denial groups. While not directly funding research, they do continue to fund the educational and political activities of denier groups. Other funders have also stepped up to the plate with increased direct and indirect support for such groups. These included the American Petroleum Institute and the Koch Brothers, major donors to conservative causes. By 2009, two groups, Donors Trust and Donors Capital Fund, became very influential front groups which, under revisions in U.S. law, were not required to reveal who their donors were, thus assuring anonymity to wealthy funders. Rather than directly funding climate denial efforts, money could now be funneled secretly through such front groups. These two groups alone soon accounted for 25 percent of all funding for climate change deniers. (19) Charles G. Koch and David H. Koch are prime examples of where the denial money comes from. They have a vested interest in delaying climate policy action. They have made billions from their ownership and control of Koch Industries, an oil corporation that is the second largest privately held company in America, which also just happens to have an especially poor environmental record. It is conservatively estimated that the Koch Brothers have funneled $67 million to climate-denial front groups that are working to delay policies and regulations aimed at stopping global warming. (19) But this dollar amount is no doubt vastly underestimated.

As we have noted, the amounts that the Kochs and others are contributing are getting harder and harder to ascertain. Donors Trust and Donors Capital are not, as we have said, required under American law to reveal the name of their contributors. This makes it possible for groups and individuals like the Kochs to publicly back off of direct funding of climate change denial and to funnel additional monies secretly. These two groups have made it possible for anonymous billionaires to provide at least $120 million to over 100 groups between 2002 and 2010. Even this is a small portion of the millions invested to attack the science, create public doubt, and build a vast network of think tanks and activist groups working toward a single purpose: to redefine climate change from a neutral scientifically established fact into a highly polarizing "wedge issue" for hardcore conservatives. (20) As the anonymous billionaires and their front groups pushed ahead, the science has gotten only more settled. The UN's Intergovernmental Panel on Climate Change (IPCC) issued its fourth major assessment in 2007. This report, assessing the peer-reviewed science, confirmed that climate change is occurring, and most importantly, mostly as a result of human activities. With more than 500 authors and 2,000 scientists from over 60 nations offering expert review, this report was regarded as the most definitive and accurate assessment to that date on global warming. It concluded that the science had demonstrated that the warming of the climate system was unequivocal. It also concluded that the level of confidence had increased to such a degree that it was now possible to say that the global average net effect of human activities on the climate since 1750 has been one of warming. (21)

As one might expect, the 2007 IPCC report led to another escalation of efforts to discredit the science. In April of 2008, Americans for Prosperity (another Koch Brothers–funded free enterprise front group) launched a "hot air" tour to oppose carbon regulation and to discredit the science about global warming. Saying its goal was to educate American citizens, the thrust of the effort was to convince the American public that global warming and the science that supports it are a part of an alarmist and dangerous political agenda. To pursue it would result in lost jobs, higher taxes, and less freedom. This grassroots effort was, of course, a part of the ongoing political strategy to create public doubt and win a policy debate.

With the election of Barack Obama to the presidency, some expected a more proactive policy approach to the climate challenge. Indeed, in his first State of the Union address in 2009, President Obama spoke of the need to address the "ravages of climate change." He called for a market-based cap on carbon emissions. This would have capped the overall level of carbon emissions that could be produced. Companies that exceeded their specific emissions cap could, under this approach, lease additional emission credits from companies that produced less than their allotted amount. This cap and trade approach was a market-based approach to create financial incentives to reduce carbon emissions. Predictably, the proposal went nowhere. (17) Just as predictably, the summer of 2009 saw corporate efforts to discredit climate science heat up. The Heartland Institute, another noted conservative "free-enterprise" group dedicated to conservative causes and policies, published a remarkable report even for groups dedicated to promulgating skepticism about the science that suggests humans are warming the planet. They noted that the earth may indeed be warming, but the cause was, contrary to all scientific evidence, entirely natural. This report even concluded, again contrary to all scientific findings, that overall, "a warmer world will be safer and healthier for humans and for wildlife." (17) In the summer of 2009, Congress did come close to legislating something to regulate carbon emissions, but in the end that failed also.

The Waxman-Markey carbon regulation bill, better known as the American Clean Energy and Security Act of 2009, passed the House by a vote of 219–212. The bill was intended to reduce pollution from global warming and move the United States toward a clean energy policy. This included, for example, setting performance standards for coal-fueled power plants and providing funding for training workers for jobs in the renewable energy industry. But the bill never got a vote in the Senate. Democratic Senate majority leader Harry Reid refused to bring the American Clean Energy and Security Act forward for a vote. He said he simply did not have the votes (60 were required under Senate rules) to pass it. The collapse of this legislative effort was a major setback for any reasonable chance to act on climate change in 2009. (17)

By October of 2009, as the political debate raged and the actual science became muted by the efforts to distort and misrepresent it, public opinion reflected the manufactured political disagreements more than the scientific consensus. A Pew Research Center poll found that only 36 percent of Americans believed there was solid evidence that the earth was warming because of human activity. This was down from 47 percent in 2008. More astounding, in late 2009, only 57 percent of Americans believed that there was solid evidence that the climate was warming at all. This compared to 71 percent in 2008. (17) Clearly the back and forth between the science and the not-science forces was creating public doubt even as the science was becoming more settled and certain. The year 2009 ended with a manufactured and entirely false "controversy" that fueled the doubts being promoted by deniers and funded by anonymous billionaires.

In late November 2009, just before delegations were preparing to assemble in Copenhagen for a global climate change summit, more than 1,000 emails between scientists at the Climate Research Unit of the U.K.'s University of East Anglia were stolen and made public by unnamed hackers. Climate skeptics claimed that these hacked emails showed scientific misconduct that amounted to the complete fabrication of man-made global warming. (17) Such allegations were investigated by the IPCC and discovered to be entirely unfounded. The messages, which spanned 13 years, showed a few scientists in a bad light being rude or dismissive of legitimate skeptics. They also showed some legitimate discussions or exchanges about methodological adjustments to improve climate modeling or attain more accurate measures of climate-change impacts. But none of the emails were found to have improperly affected the actual scientific work or the consensus on climate change or to cast any doubt on the scientific evidence that the earth is getting warmer and that humans are largely responsible. A few skeptics and all deniers claimed this trove of emails showed that the scientists at the U.K. research center were engaging in evidence tampering, and they portrayed the matter as a major scandal: "Climate-gate." (17) The news coverage of this faux scandal only served to create the illusion that the science was unsettled and that scientists were engaged in some sort of struggle to manipulate data or present pure fabrications as scientific fact. But almost unnoticed by some, the legitimate science was becoming more settled than ever.

As 2010 began, NASA reported that the decade 2000–2009 was the warmest on record. The National Academy of Sciences released its most comprehensive climate report up to that date. It stated unequivocally, based on a summation of reliable, peer-reviewed science, that climate change was occurring, that it was caused largely by human activities, and that it posed significant risks for humanity and for a broad range of natural systems. (17) Politically, the issue of climate change was little mentioned during the 2010 midterm elections. But the results of those elections, a sweeping victory for Republicans, including the first wave of Tea Party candidates, had implications for climate politics. A new Republican majority in the U.S. House of Representatives (most of whom stridently rejected the findings of science and opposed action on climate change) quickly voted to eliminate the House Committee on Global Warming. Saying this was done in an effort to eliminate waste, new Speaker of the House John Boehner said that climate change would henceforth be handled by the science committee. (17)

In January of 2011, President Obama spoke about energy but made no mention of climate change in his State of the Union address. In March, Congressman Henry Waxman, a Democrat from California, proposed several amendments attached to legislation that in effect said simply that "Congress accepts the scientific finding of the Environmental Protection Agency that warming of the climate system is unequivocal." This language was rejected in committee each time it was proposed, with all of the Republican members of the House Energy Committee voting against it. (17) The science has been settled for some time, but as the science has become more and more settled, the politics has become more and more unsettled. Those elements in our public discourse that have sought to create doubt about the science, who have in effect waged a long campaign funded by invisible billionaires and fossil-fuel interests, have created out of scientific unity and consensus such manufactured political dissonance as to make climate change one of the leading issues dividing the public along partisan lines.

In December 2011, a Pew Research Center Poll confirmed that public doubt remained stubbornly unaffected by climate science. Among those polled, 38 percent believed that human activity is the main cause of global warming. This was a modest increase over the 34 percent figure for 2010. But far more interesting in relation to the discussion of this chapter, the partisan divide in public opinion was increasingly stark. Over half (51 percent) of Democrats and 40 percent of independents agreed with the scientific consensus that climate change was due primarily to human activity. Only 19 percent of Republicans and 11 percent of Tea Party Republicans agreed with the scientific consensus. (17) It is certainly interesting (also expected based on the way the politics of climate change have played out) to note the incredible disconnect between scientific and public opinion. This disconnect may actually have become so complete that the mere discussion of climate change in public and political discourse is deemed as a liability in attracting support and winning elections. This seemed to be the case in 2012, at any rate.

According to the U.S. Climate Change Science Program, 2012 found the United States experiencing more unusually hot days and nights, fewer unusually cold days and nights, and fewer frost days. Storms and severe weather events were becoming more frequent and intense. Weather extremes were becoming more frequent. In January of 2013, the National Climate Data Center in Asheville, North Carolina, reported that 2012 had been the warmest year on record in the United States. The year 2012 also included 11 natural disasters in the United States that reached the $1 billion threshold in damages and losses. This culminated with Superstorm Sandy in October and its $65 billion in damages. With the sheer number of recurring and expected natural disasters and their costs, with the ever-more-precise and widely accepted science linking climate change to forecasts for future extreme weather events, inaction by our elected officials is hard to explain. With constantly and rapidly changing risk and vulnerability profiles associated with a changing climate, and with the implications all of this may have for the resilience and sustainability of our communities across the country, it is difficult to imagine that politicians and elected officials would not be engaged in meaningful dialogue and discussion about climate change. Yet, in the 2012 national election in the United States, the issue of climate change was not seriously discussed as a major concern. One of the national parties ridiculed the science, and the other seemed eager to say as little as possible about it in fear, no doubt, that too much discussion would prove costly on Election Day in a nation deeply divided over the issue. The consensus of the scientific community, the mounting evidence suggesting that climate change may be the most important challenge facing humanity in the 21st century, and the objective facts of the matter had little influence on the political agenda or the conversation of 2012.

In the four major presidential debates between President Barack Obama and Republican nominee Mitt Romney, the issue of climate change never came up. Not a single question about it was posed by debate moderators, and neither candidate brought it up. Throughout the entire campaign, the president and Mr. Romney worked hard to outdo each other as lovers of coal, oil, and natural gas. These, of course, are the very fuels that contribute most to the rising levels of carbon dioxide in the atmosphere. Neither presidential candidate, nor either of their political parties, for that matter, raised or addressed any of the most important and critical economic, environmental, political, and humanitarian issues a changing climate will force upon the planet and the people who occupy it.

Among the unasked questions, a few seem quite basic and quite important. Should the United States cut its greenhouse gas emissions? If yes, how far and how fast should we reduce emissions? Should fossil fuels be more heavily taxed? Should any form of clean energy be subsidized, and for how long? Should the United States lead international mitigation efforts? Should the nation pour billions of new dollars into basic energy research? Is the scientific evidence serious and reliable enough to suggest that we should have a sense of urgency about acting on climate change? Is the climate system so fraught with uncertainty that the rational response is to do nothing? Could it not at least be suggested that the science and the scientific consensus had evolved to the point that such questions should logically have been raised and a discussion of them an important national concern in 2012? It is difficult in the context of what the science has shown us to reconcile oneself with the fact that such questions were not raised and discussed during a national election.

Many scientists and policy experts felt that the lack of a serious discussion of climate change in the 2012 presidential contest represented a lost opportunity to engage the public. It also represented a lost opportunity to signal to the rest of the world American intentions for dealing with what is, by definition, a global problem that requires global cooperation. There are perhaps some signs that our national discussion has the potential to become a little more elevated and serious, but many of the partisan and political impediments to more meaningful policy action remain. Indeed, the issue was once again not much discussed in the 2014 midterm elections in the United States. The Republican gains in the House in 2014, increasing their majority, and a new Republican majority in the Senate, locks in place for the foreseeable future a strong, uncompromising Congressional majority that will refuse to acknowledge the existence of a climate crisis and that will oppose most policy measures that address climate change. It is not unreasonable to ask, given the decades-long and ongoing debate, what is the reaction of the American public to all of this? It is in fact very important to know what the American people think, why they think it, and what difference what they think makes.

Public Opinion and Climate Change

Public opinion on climate change is subject to the same variety of influences that shape public perceptions on any issue generally. Beyond the voicing of opinions in elections, opinions that are subject to the influences of political advertisements and the partisan framing of campaign issues, the general opinions of the American public may expand or limit the options that policy makers, once elected, consider to be politically viable. This is especially the case with respect to climate policy which, we can say, may require some lifestyle changes. Low public concern or mixed opinion about climate change, for example, may be seen to impede policy progress. That seems to have been the case for the past several decades, and more to the point of addressing the climate crisis, would appear to be one of the major things that must change if we are to respond sensibly and efficiently to the climate crisis and what may very well be a looming climate disaster. The general public must be transformed from obstacles into enablers of climate policy initiatives that can address the challenges that reliable science tells us are posed by a warming climate.

As we consider the role of public opinion in relation to climate change, it will be more apparent that a part of the required strategy to respond to or manage what we are calling the climate crisis must include a reframing of climate change. It must be persuasively articulated as a national and global concern in such a way as to move public perceptions to that place where they are no longer impediments to policy action but are informed sufficiently to enable progress. This means broad public acceptance of the fact that we have already reached the crisis point, that is, a critical point of decision which, if not handled in an appropriate and timely manner, or if not handled at all, may turn into a disaster or catastrophe of unprecedented proportions. But just how far away are we from reaching that point?

In February of 2013, Orie Kristel of The Strategy Team, an applied social science company based in Columbus, Ohio, provided an interesting overview of public opinion about climate change in an interview published in *USA Today*. (22) The major conclusion of his public opinion research is that the public belief that global warming is happening has been mostly stable and increasing over the past 30 years.

The question "Do you think the greenhouse effect really exists or not?" was first asked of U.S. respondents in 1986. At that time, 73 percent said yes. Over the years, when pollsters have asked people whether they "believed" climate change was happening, a similar majority have agreed that it is. However, when asked by pollsters if there is a solid scientific consensus that the average temperature of the earth has gotten warmer due to human influences over the past four decades, the percentages in agreement were considerably lower. (22) What does this suggest? Apparently, if you ask people if they believe global warming is happening, you get one answer (i.e., over 70 percent agree), and if you ask them to assess the scientific consensus or what they know about it, which in many cases is very little, you may get another (i.e., closer to 50 percent). A majority still agree in the latter case that science has proven man-made global warming is happening, but it drops considerably below 70 percent. All of this is to say that the question one asks, the way one asks the question, or the way the issue is framed in the survey instrument, may influence the responses one will get. Terminology or language choices might matter. Kristel and his colleagues weighed together public opinion polls, dating back to 1986, from more than 150 nationwide questionnaires.

Belief that global warming is happening has approached 75 percent. It dipped a bit in recent years, but the public consensus has since resumed its upward march. Kristel and his Strategy Team colleagues report that the wording of poll questions does seem, in some cases, to have created some of the appearance of shifts in public opinion about global warming. (22)

While it does appear that more specific questions about the status of climate science, or the peer-reviewed conclusions within the scientific community, may produce less public awareness of and/or agreement with the scientific consensus, primarily because fewer people actually have knowledge about the science or the consensus, it does not appear to follow that the choice of terminology itself actually exerts a major influence over public opinion. In a 2011 article, two scholars explored the terminology question in some detail. (23) They raised the question of whether the term *climate change* versus the term *global warming* affected the perceptions of respondents. Likewise, they questioned whether the use of different terminology in describing the cost of climate change mitigation might influence opinions. For example, did defining mitigation costs in terms of "higher taxes" or "higher prices" have an impact on levels of support for mitigation efforts? Political strategists, of course, are always looking for terminology that will produce a specific response in public opinion. Many of these strategists have sought to determine for over the past decade if the use of the terms *climate change* or *global warming* would impact public perceptions. Their goal, of course, is to find language or terminology that will influence public opinions in the direction favorable to their causes and candidates. This study, published in 2011 in the journal *Climate Change*, concluded that the choice of terminology, as opposed to the wording of a question in a survey, may not influence opinion much at all. (23)

Of greater interest perhaps than the number of people in surveys who generally believe that climate change or global warming is happening is the number of people who believe that the government should do something about it. A study completed in 2013 by scholars at Stanford University demonstrated that from 2006 to 2012, a large majority of Americans supported the idea that the government should require or encourage policies reducing greenhouse gas emissions. This included restricting the amount of greenhouse gases that businesses are allowed to emit, requirements for more fuel-efficient cars, building more energy-efficient buildings, and a variety of other measures. Seventy percent or more supported each of these options. (24)

The policy with the lowest level of public support was building electric vehicles. But a majority still supported that.

Another interesting question asked in the Stanford survey was whether the government should require policies designed to reduce greenhouse gas emissions. When these policies were not framed or presented in terms of taxes or costs, the response was overwhelmingly yes. When specifically asked if they would support giving tax breaks to produce electricity from water, wind, and solar power, large majorities of Americans supported this policy from 2006 to 2012. A majority also supported giving tax breaks to reduce air pollution from burning coal. When it comes to giving tax breaks to build nuclear power plants, support drops below 50 percent. As for increasing taxes on gasoline and increasing taxes on electricity, a large majority of Americans did not support or endorse these policies. This would seem to suggest, in other words, that proposed increase in taxes to pay for climate change mitigation efforts reduces support for policy initiatives. (24)

That the American public believes government should do something but that it does not want to pay for it is not at all unusual. This is not an atypical result with respect to most issues. Likewise, as with respect to most issues, public opinion is often more complex a phenomenon than the numbers reported in an opinion survey may reveal. But some surveys do reveal more than others. Efforts to get at the intensity of public perceptions and/or to quantify multiple and identifiable levels or gradients of opinion often provide a fuller picture than the snapshot feeling thermometer or issue poll. The Yale Project on Climate Change Communication has revealed, for example, that public opinion about climate change, like all issues, is multilayered and that the "public" is really a plural concept, that is, we are talking about "publics" and not a single monolithic public.

Researchers at the Yale Project on Climate Change Communication have identified six unique population segments that perceive and respond to the issue of climate change in distinct and different ways. In surveys conducted in August and September of 2012, these Six Americas were found to cover a wide spectrum of concern and issue engagement. The extreme ends of the spectrum consist of the segments that accept and reject climate science with several publics in between. (25) Table 3.1 shows the breakdown of the six publics. Climate science, or the peer-reviewed research in the field, when we examine these publics in some detail will be seen to have had minimal impact on the thinking of four of these six publics. With respect to

Table 3.1   Global Warming's Six Americas

1.	ALARMED	16 percent
2.	CONCERNED	26 percent
3.	CAUTIOUS	25 percent
4.	DISENGAGED	5 percent
5.	DOUBTFUL	15 percent
6.	DISMISSIVE	13 percent

Source: Yale Project on Climate Change Communication http://environment.yale.edu/climate communication/files/Global_Warmings_Six_Americas_book_chapter_2014.pdf.

the other two, the two publics that have some greater awareness of what the science is saying perhaps, it appears to have produced strong and different reactions. One of these publics accepts the science and the other rejects it entirely.

It is worth summarizing the findings of the 2012 Yale study in a bit more detail.

The "Alarmed Public" (16 percent) consists of those who are very concerned about climate change and support aggressive action to meet its challenges. This group would be what political scientists call an issue public. This means that they have studied the issue and gathered information, they have a desire for more information, and they have a sense of urgency about acting. In the case of climate change, they have more information about it than the average citizen and strongly agree with the scientific consensus. Almost all in the "Alarmed Public" are certain that global warming is happening and believe it poses real risks to them today and to the generations to come. More than 75 percent or three-quarters of this public agrees with the scientific consensus that global warming is primarily human caused. For this "Alarmed Public," global warming requires a sense of urgency with respect to public policy, etc. (25) The Alarmed do have a higher proportion of Democrats, and about 50 percent identify themselves as liberal on the ideological scale. But almost half of this group does not identify as liberal. The "Alarmed Public" is also better educated than the national average. (6)

The "Concerned Public" (26 percent) is less likely to see global warming as a threat to themselves or their communities today. But to the extent that they do perceive it as a threat, they see it as far off in the future and not of immediate concern. The "Concerned Public" accepts that global warming is happening, but its perceptions about the issue differ from the "Alarmed Public" in several significant ways. They are less likely to see global warming as linked to human causes, less likely to feel that any of the risks associated with it are of immediate concern, and are less likely to know about or accept that there is a scientific consensus. Yet, with respect to each of these items, small majorities do agree with the "Alarmed Public" on key beliefs and perceptions. The greatest difference between the "Concerned Public" and the "Alarmed Public" is that the concerned segment has not thought about the issue as much, has lower levels of information, and almost half (compared to 18 percent of the Alarmed segment) say they do not need or want more information about the issue. (25) The "Concerned Public" is more middle-of-the-road politically. Its liberal and Democratic leanings are only slightly higher than the national average. With respect to demographic characteristics (gender, age, ethnicity, income, education), the "Concerned Public" is reflective of the national average. (25)

The "Cautious Public" (25 percent) is weak on all opinions or beliefs associated with climate change. They are more likely than not to believe that the climate is warming, but they are unsure what to make of it. Only 19 percent are "certain" that the climate is warming. Eight in ten perceive it as a possible risk for future generations and of little or no immediate concern. Fewer than half agree that global warming is human caused, just over 40 percent think that the science is settled or that the scientists have achieved any consensus, about a quarter believe the United States is currently being impacted by global warming, and only one percent say they have given the issue much thought. (25) The "Cautious Public" is less likely to be college educated than the Alarmed or Concerned segments but are right on the national average with respect to ethnicity, gender, race, age, and income. They are inclined to be suspicious of claims that climate change is human caused, less informed about the science, and 70 percent say they pay little to no attention to climate change information or news. (6) The "Cautious" are unsure, uninformed, and not paying close attention. This makes this "public" more uncertain and unimpressed with science. Of all six "publics" identified in the Yale study, the "Cautious" and the "Disengaged" have the lowest attention levels with respect to environmental news generally. (25)

The "Disengaged Public" (5 percent) is the one that has given global warming the least amount of thought. "Don't know," where that is an option, is their preferred response to all questions about the issue. For example, 98 percent said they "don't know" if climate change will be a great challenge to future generations. Only six percent of the "Disengaged Public" thinks global warming is happening. The "Disengaged Population" are the least likely to have graduated from college, they have the lowest incomes, and they are more likely than other publics to be poor, retired, or disabled. (25) They tend to be politically inactive (moderate Democrats who do not vote) with the lowest number of registered voters and the lowest levels of interest in politics. Interestingly, they are more likely than average to believe in biblical literalism and to reject the theory of evolution. Eighty percent say they have paid little to no attention to any global warming information. (25)

The "Doubtful Public" (15 percent) is similar to the "Concerned Public" with respect to interest and involvement with the issue of global warming but display a much lower level of acceptance of any of the scientific conclusions about the subject. The "Doubtful" are inclined to disbelieve that global warming is happening (only about 40 percent agree that it is). They are even less inclined to see it as a serious problem. Only 22 percent believe that it may present risks of any kind to future generations. Only 12 percent agree that global warming is linked to human causation. Only 13 percent believe that the scientists agree or have reached a scientific consensus about the fact, causes, and effects of global warming. The doubtful do not see how the scientists can actually know that climate change is real or that it is human caused (i.e., the proportion of the doubtful who ask "How can they know?" is the highest of any of the six publics). This doubt is difficult to address, as only three percent of the "Doubtful Population" say they pay much attention to global warming information. (25) The Doubtful tend to be politically conservative (over half), and most identify as Republican. They are also more likely to be white, male, and conservative than the national average. With respect to education, income, and age, they are close to the national average. (25)

The "Dismissive Public" (13 percent) is the most certain that climate change is not happening and the most confident in their opinions. These are the climate change deniers, and they are the least likely to accept any information, scientific or other, that contradicts their beliefs. None (zero percent) believe that global warming is harming the planet now, and only eight percent think it is happening at all. Only six percent say the scientists are in agreement or a consensus has been reached about climate change. The "Dismissive Population" is firm in its rejection of climate science and its conclusions and the most likely population to say that it does not need more information to make up its mind on the subject. More than 70 percent of the "Dismissive Population" are conservative, and over 50 percent are registered Republicans (only 3 percent Democrats). They are more likely to be white and male than the other populations, and their socioeconomic status is higher. The Dismissive have greater educational attainment and the highest income of the six Americas. The Dismissive are also the least likely to trust scientists on climate research and the segment most likely to follow politics closely. (25)

As our brief summary of the "Six Publics" suggests, and as the authors of the study demonstrate in their analysis, the American public is a diverse population. Communication about climate change, or any issue, for that matter, cannot treat the "public" as a homogeneous mass. Diversity of opinions, the cultural and political underpinnings of those opinions, and differing information levels and interest levels of population subgroups complicate communication and make the challenge of mobilizing public opinion to support significant climate policy very difficult. Politicians, the media, and governments have all developed the art of selecting target audiences and tailoring communication to them. This is most effectively done to create or enhance already existing divisions rather than to produce unity; at least in our recent political experience, that is very much the case.

A third section of the Yale study revealed how each of the Six Americas perceived the relative influence of individuals, organizations, and companies on the elected representatives who are shaping U.S. energy and climate policies. All six publics believed that they, and the people who shared their opinions, had less influence on public policy than campaign contributors, fossil-fuel companies, and the media. Each public, in fact, perceived itself to have the least influence. Five of the six publics believed that large campaign contributors had the strongest influence on public officials. Four publics (Alarmed, Concerned, Cautious, and Disengaged) believed that the fossil-fuel industry had more influence than the renewable energy industry. Two (the Doubtful and the Dismissive) felt that the renewable energy interests had more influence than the fossil-fuel interests. The Dismissive population predictably said that the "liberal news media" had the greatest influence on elected officials. (25)

As the Yale study conducted in 2012 suggests, and as subsequent studies continue to show, we, the American people, do not think clearly about climate change. In March of 2014, the International Panel on Climate Change (IPCC), in its fifth report, has stated in the strongest terms yet that global warming is occurring and humans are to blame. (26) As noted in a 2013 study by the Yale Project on Climate Change Communication, there is near unanimous agreement from climatologists about anthropogenic, or human-caused, climate change. In other words, there is a near-unanimous consensus about this among the scientists conducting the peer-reviewed research in this field. Yet the majority of Americans (six out of ten in a 2013 sampling) are unaware of this consensus. (27) Other studies indicate that whatever their opinions about it, climate change is not regarded as an important or urgent concern by far too many people. (28)

In a report issued in early 2013 at the beginning of President Obama's second term, the Pew Center for People and the Press detailed the issues that Americans viewed as the "top priority for the president and Congress." The economy and jobs dominated the list for the fourth straight year—not a surprise, as the effects of the 2008 economic crash continued to be felt—and the national debt was a close third. Out of 21 possible priorities identified, protecting the environment ranked 12th, and dealing with global warming ranked dead last. Also, as indicated by other studies, public responses followed somewhat along partisan lines. (28)

Political partisans and special interests have taken increasingly polarized positions on climate change. There have always been partisan differences over environmental issues, of course, but a recent study suggests that partisan warfare over climate change may have reached a new level in stoking existing divisions in the public at large. (29) This is to say that partisan political conflicts affect what everyday Americans think about the threats associated with a warming climate. A study appearing in a 2012 issue of the journal *Climate Change* combined the results from all available surveys between 2002 and 2011 to create an index that measured the level of public concern about climate threats every three months. In the assessment of factors that influence public thinking about the subject, partisan political attitudes were found to have the major impact. (29) It may be useful to examine briefly what this study found to be the drivers of public concern about climate change.

Scientific information and unusual weather were two variables found *not* to have as much of an impact in moving public opinion about climate change as one might wish to think. Most of the public does not read even routine scientific analyses, and these have zero impact on public opinion about climate. The release of major reports and articles on climate in popular magazines has only a slight impact.

Media coverage was found to have a small effect. The changing frequencies of stories highlighting climate-change impacts did have a significant though temporary effect on levels of public concern. The analysis also showed that the activities of advocacy groups that spend time and money to mold public opinion have only a modest effect on public opinion. Overriding and more immediate events such as war or economic recession invariably contribute to significant declines in public concern about climate change. But it was the effect of partisan sentiments, party identifications, and ideology that was found to be the largest single factor in explaining the ups and downs of public opinions about the threat of climate change. (29) In other words, Republicans and Democrats have taken polarizing positions on climate change and use their media megaphones to work up their respective bases. This has influenced what the public thinks about climate change such that the distribution of public attitudes falls increasingly along partisan lines. In the spring of 2014, a midterm election year in the United States, public concerns about climate change as measured in a variety of surveys rated near the bottom. Even where respondents might be concerned about climate change, it continued to rate at or near the bottom of the list of priorities for policy action. When results are compared for partisan differences, it is of course no surprise that identifiers of the two major political parties express very different levels of concern about different issues. Climate change is no exception. As has been the case since 2008, the major concerns on the minds of the electorate related to the economy. The level of concern about climate change remains stubbornly low. Again, in the midst of more immediate concerns or crises (the economy, war) climate change fades as an issue of concern. Be that as it may, Democrats consistently display more interest and concern than do Republicans about it.

Given the apparent low priority assigned to climate change as an issue in recent measures of public opinion, and given the conflicting and contentious messages the public receives from partisan and economic elites about it, public opinion on climate change is very likely to remain divided and unscientific. It appears to be shaped more by the mobilizing efforts of ideological advocacy groups and elites than by any accurate scientific information. (29) But beyond the findings of any particular public opinion measurement—and we must remember that such surveys are snapshots of a frozen moment in time and subject to change—there may be a few more basic things that inevitably influence not only what people think but why they think it. It is to a discussion of what I am calling some of these "basic things" that we now turn our attention as we examine more broadly why people believe what they believe. Beliefs, which are frequently the basis for opinions, it turns out, are not usually related to scientific fact. That handicaps discussion and communication about climate change in important ways. It increases the level of difficulty in no small measure.

Why People Believe What They Believe

Based on our discussion of the six publics in relation to public attitudes about climate change, it can be said that nearly seven in ten Americans (69 percent) are at varying levels disengaged from the facts about climate change. (25) Their levels of knowledge and interest, as they decline, separate them from the science and render their opinions subject to other influences. It is also apparent that roughly half of us do not spend much if any time thinking about climate change. (25) Yet we do have opinions. The absence of knowledge or unfamiliarity with the science does not mean that Americans do not have opinions or beliefs about global warming. We might benefit by taking a moment to reflect on how we generally come to have an opinion or a belief that influences our opinion about global warming despite our often being disengaged from fact and science.

While the effects of extreme weather events are short-lived and not long-term influences on what the public thinks about climate change (29), it is nevertheless true that what people generally believe often does seem to shift with the seasons. After a bitterly cold winter, belief in the reality or concerns about the seriousness of global warming will decrease slightly. After an extremely warm summer, belief and concerns will increase slightly. (29) During the brutal winters of 2014 and 2015 experienced in parts of the United States, it was not unusual to see shivering pundits on the cable news channels (at least one of them) and senators with snowballs scoffing at the whole notion of climate change. It is stunning, in light of the time and effort science has taken to document and explain the nature of climate change, to see that "beliefs" about it can shift so quickly and so thoughtlessly. When the weather is hot, especially if it is unusually so, the level of public concern about the reality of global warming seems to increase. When it's cold, some people shrug off their previous concerns, and "belief" decreases. Weather-related acceptance or denial is troublesome to the scientists and problematic with respect to the public policy dialogue. (30)

As the cold and stormy winters of 2014 and 2015 unfolded in the Eastern United States, scientific researchers did point out that, despite the short-term and brutal cold in some places, winters (speaking globally) have been getting warmer on average. While the midsections of the United States experienced a harsh winter in 2014, California and the Southwest were in the grips of record drought. Australia was experiencing (for them) a brutal summer heat wave at the same time. According to NASA researchers, one of the reasons bitterly cold winters and global warming coexist is a phenomenon known as Arctic Oscillation. This refers to the interaction of the jet stream and Arctic air during the winter. It can cause, due to increased warming in the arctic, unseasonably cold air masses to sweep down over what are normally more temperate latitudes, making for unusually cold temperatures and severe winter storms across many parts of the United States, for example. (30)

In another sense, the notion that "it was cold this winter, so the climate cannot be warming" is a misunderstanding of trends. I am referring to the inability to differentiate between a long-term trend and a short-term wobble or variation within that trend. People notice this winter's or this summer's weather but not ten-year climate averages or 50-year trends. The inability to distinguish between a trend and a short-term variation within a trend is to be frequently observed in public thinking. This inability is, sadly, rooted in a fundamental scientific ignorance. This is not to be critical. It is only to suggest that too many people, even in this advanced and highly technological age, do not understand the basics of science or the scientific method. In other words, *climate change is complicated*, and often our thinking about it is just too uninformed or too simple. Hence our beliefs about it are subject to irrational swings. Some of these swings are swayed by weather. Weather is not climate (though it may be influenced by climate), but it is immediate for people to notice and to base an opinion on, even if the information gained from it is not particularly accurate or relevant in relation to longer-range climate trends.

It is not that the basic concept of global warming is too complex to understand. It is the details that are not always easy to understand. They can, in fact, be confusing. Some places will get warmer, and some will get colder. Some will get wetter, and some will be drier. It is not easy for people to understand or see exactly what will happen, when exactly it will happen, and where exactly it will happen. Understanding climate change simply requires more effort and time than many people are willing or able to give it. Most people do not

have the time to sort out all of the intricacies of global warming because they have very busy lives. But somehow, if we are ever going to
be able to act with any sense of urgency and intelligence in the public
policy arena, it is imperative that the public come to understand four
things that are both true and relatively simple in the end. One, global
warming or climate change is really happening. Two, this is a bad
thing. Three, the global warming we are experiencing is caused by
humans. Four, there are things we can do and need to do about it.

Among the things that may frustrate the ability to achieve a public
sense of urgency about the need to act in response to global warming—and remember, most people concede it is happening, even if they
are divided about its causes or uninformed about the science—is
what might be called the *temporal distortion phenomenon.* The speed
or timing of global warming is slow and gradual, though not as slow
and gradual as we may perceive. It is happening right now, and the
need to act is urgent and immediate. To the average citizen focused
on the weather outside the window this morning, scientific information that tells them that global temperatures and global sea levels are
rising, but at what seem to them to be tiny fractions from year to
year, seems insignificant. While the trend over the next decade or
several decades may suggest something terrible unfolding in the future, it is difficult to worry about that when it all seems so irrelevant
to our short-term focus. Even when everything is expertly explained
and quantified, and the vulnerabilities and disasters awaiting us in
the future are made clear, human beings are not very good at caring
about tomorrow.

Temporal distortion is enhanced by the fact that American culture
is notoriously shortsighted. Shortsightedness is the human norm, to
some extent. We never, as a habit, focus on the long haul. Our tendency to focus on immediate events combined with a short attention
span makes the current situation (whatever it is today) more influential on our thinking than the future. Public opinion is fickle. Changing
circumstances, the debate amongst political elites, and changing media coverage may all influence it dramatically. Climate change is an
issue where this is a prime example. American public opinion has
bounced back and forth on this topic over the past decade. Between
2007 and 2010, there was a sharp decline in the number of Americans
who believed that climate change was real and that it needed to be
addressed as a policy priority. That decline was attributed to the bad
economy (the worry about jobs pushed climate and environment to a
lower priority), the debate among partisans (contributing to the

doubt about the science or perception of the threat as exaggerated), and to media coverage (or the lack of it). By the summer of 2012, public opinion polls showed a sharp increase in the number of Americans who believed that climate change was real and that policy makers should address it. The record summer heat and persistent drought experienced across much of the country was undoubtedly having an impact on public perceptions.

The fact of the matter is that immediate concern, the crisis de jour, if you will, always seems to carry more weight in public thinking and conversations than longer-term concerns or objectives. Now, as in today, is never the right time to think long term or to make changes in the way we live. There is always an accepted, usually false, excuse for not acting to address long-term concerns. Fixing a stalled economy must take immediate and urgent priority over policies to address the impacts of global warming, and our current energy needs are said to dictate continued reliance on fossil fuels. The alternatives are falsely believed to be not available, not affordable, or not practical, etc. (in the short term), and of course the costs associated with sustainability enhancing practices and policies to address global warming are (always) declared prohibitive at the present time. If not that, policies designed to address climate change are said to be harmful to the economy in general. They are not, especially if we are intelligent about it, but that's the argument made by those with the largest stake in the status quo and the greatest reluctance to accept and deal with the reality of global warming. The accepted reasons for not changing course and the often-unquestioning acceptance of them seem most frequently to prevail. Generally, this is to our detriment and far more costly to us in the long run. But we do not live in the long run, or so it seems sadly reasonable to conclude.

An assumption that the crisis de jour, whatever it may be, cannot be addressed while at the same time addressing longer-term threats like global warming is the common by-product of short-term thinking. Short-term thinking, in turn, escalates negative consequences. It leads to the acceptance of what we are doing now as the necessary thing to do, and not infrequently, the assumption that the people doing what they are doing are the experts and the technologies that support their activities can be managed safely. These "experts" spend huge amounts of money to tell us everything they are doing is necessary, urgent, and safe. So we needn't worry, at least in the short term. We often accept as common sense the course we are on without questioning it. When the drillers for natural gas tell us that horizontal hydraulic fracturing is perfectly safe, we believe it because they have told us and they are the experts. If we don't believe it entirely, we still may be willing to roll the dice on risk a bit because we need the jobs, or we are worried about America's energy future, and these things must take priority over our concerns about the safety of horizontal gas drilling and any risks it may impose upon the environment or humans and the communities in which they live. The first wrong assumption (i.e., we have to deal with an immediate crisis first and postpone any concerns about sustainability in the long term) leads to a series of false assumptions upon which are built multiple disasters. With respect to climate change, the worst effects or impacts of which are thought to be in the future, it is too easily assumed we have time, and there is no need to panic.

The insidiousness of temporal distortion and shortsightedness is disturbing on a couple of levels. Around the globe, we continue to see new record-high temperatures and a reduced number of record lows. But the exceptions to this long-term trend, for example, "it was brutally cold this winter," breed a false complacency. It encourages some to deny that the climate is warming and some others to simply assume we have ample time to address the problem. When one adds the focus on more immediate and pressing issues such as the economy or war, it is easy to let the concerns about climate change drift and to go unaddressed. The illusion that climate change does not require an urgent or immediate response, if any at all, stubbornly persists. Coupled with the assumption that we cannot address the problems associated with global warming because we have more pressing needs (economic, energy related, other) that must come first, this illusion plants the seeds for a disastrous future few can see coming with a near or short-term focus. Humans, and especially we Americans, are simply not wired to see signs of disaster resident in their present actions or inactions.

Humans are generally tuned into extraordinary events with shocking visual imagery that happen with unexpected or unusual intensity. Dramatic storms, fires, droughts, floods, and weather extremes will capture our attention because they contrast with what we regard as normal or expected. This *contrast effect* may cause some slight shifts in perceptions or beliefs about global warming. But unless these extraordinary events happen repeatedly, consistently, and in unprecedented ways, their impact on public attitudes is short-lived. With respect to climate, the extraordinary will cause a slight bump in concern, but assuming it is not a constant or ongoing phenomenon, there is an inevitable tendency to hope and/or believe that these events are short-term fluctuations in the normal climate and not harbingers of dramatic changes or any eventual tipping point. A certain amount of wishful thinking influences public belief. This is perhaps best thought of as an understandable disliking of the nastier implications of a warming climate and a natural reaction to make it go away by latching on to the convenient belief that extraordinary events of unusual intensity are a brief deviation from normal and not indicators of a new normal. Such wishful thinking is a perfect response to minimize the upsetting effects of the contrast effect as we focus on things seen as more immediate and more important than climate change. This perpetuates the illusion that climate change is not a problem that requires an urgent or immediate policy response.

Human beings have a propensity to believe what they have always believed, to accept what they already think is true, to reject contrary beliefs, and to seek out and find information that supports the position they currently hold. This is a phenomenon called *confirmation bias*. Confirmation biases contribute to overconfidence in personal beliefs and can maintain or strengthen beliefs in the face of all contrary evidence. When searching for information, confirmation bias means that a person will actively search only for information that supports their currently held belief. It is commonly said, for example, that American liberals seek their news from MSNBC and conservatives only watch FOX. This is all about the tendency to fill up your time with material that reinforces your worldview. With the number and diversity of news sources available today, and the ability to access or engage these sources 24/7 on demand, it is entirely possible for one to arrange it so that one is never exposed to any "news" that is not presented in a manner that reinforces his/her views. This selective ingesting of "news" or information contributes more perhaps to the nurturing and perpetuating of partisan or ideological division than it does to an informed public.

We have already observed that studies have shown that political partisanship, party identifications, and ideology are the largest factors that influence public opinion about climate change. (29) In our discussion of the six publics, we saw that the "Alarmed Public" has a higher proportion of Democrats, and about 50 percent identify themselves as liberal on the ideological scale. At the other extreme, we saw that more than 70 percent of the "Dismissive Public" (i.e., climate-change deniers) are conservative, and over 50 percent are registered Republicans (only 3 percent Democrats). What is the explanation for these partisan or ideological differences about global warming?

It used to be assumed that the level of education was the most important variable that explained one's attitude about scientific matters. But a recent study demonstrates that, with respect to climate change, level of education is not a variable that explains the variation in public attitudes. The study published in 2010 found that concern about climate change increased with higher levels of education among Democrats but decreased with higher education levels among Republicans. In other words, the higher the education level of Democrats, the more they believe in global warming, and the higher the education level of Republicans, the less they believe in it. These findings have been supported by other polls as well. This tells us that the divide in public opinion has less to do with science and more to do with emotions and values. Among conservatives there is a great sense of mistrust and suspicion for what they perceive to be the liberal scientific elite. (31) But this mistrust and suspicion is not the product of anything the scientists are doing. It is the result of what conservatives fear about what the findings of science might imply in the context of public policy.

Political conservatives, of course, mistrust government. They want less government and less regulation as a general and increasingly uncompromising principle. Conservatives have no need for evidence when it comes to any matter of public policy. They already know that the response they want to every political issue or concern is no policy, no government regulation, no expenditure of public resources, etc. So they work backward from their conclusion and look for an interpretation of the issue under review that will lead to their desired answer. In the case of climate change, its impacts as proven by science may be serious enough to create a strong argument in support of governmental policy action and/or regulation. Conservatives perhaps worry they cannot win an argument against taking action if they admit global warming is occurring. To prevent action (i.e., to reduce governmental initiative, which they see as always harmful), they must deny climate change's existence or at least question its seriousness as a problem. Much of the policy prescription for climate protection (e.g., carbon taxes, cap and trade) is viewed by conservatives to be anathema to free markets. The conservative response is to just say no. But to win the argument, to succeed at saying no, climate-change denial is a necessary strategic weapon.

Whatever their personal beliefs, and increasingly Republican elites are in fact turning into climate change deniers, the uncompromising abandonment of science by the conservative base and the opposition to congressional climate change legislation (led by energy companies and corporate billionaires who fund their campaigns) makes it almost impossible for any Republican who is not a climate change denier to be nominated and elected. That cannot be a good thing for anyone hoping to see reasonable policy outcomes that will address climate change realities. Conversely, political liberals are more comfortable with government in general, see a necessary and proactive role for it in relation to environmental concerns, and, based on the science (which they accept), are not threatened by any discussions of climate policy or regulatory measures.

Religious fundamentalism, like political partisanship, is associated with attitudes about climate change. Researchers have shown evidence for increasing opposition by biblical literalists to the teachings of science where it may introduce facts that conflict with their views on supposedly moral issues. Speaking generally, fundamentalists are less likely to have confidence in science and scientists than their mainstream brethren or the nonreligious. Conversely, respondents who believe that human evolution is "true" are more supportive of funding for scientific research than those who view evolution as "false" or "don't know." (32) With respect to climate change, there seems to be a correlation between religious fundamentalism and climate change denial. It would appear, for example, that evolution deniers are also climate change deniers. (33) This is not to disparage anyone's beliefs, but it is to suggest that fundamental religious belief and science do not always see eye to eye. The former, in fact, views the latter with great suspicion and is at the ready to deny its conclusions in the name of faith. As such, religious fundamentalism cannot be ignored as a huge influence on some beliefs or opinions about climate change.

Two other general influences on what we might believe about climate change should be mentioned, even though they may already be implied by our conversation of previous influences. One is the *bigness syndrome*, and the other I will label as *general cynicism*. The bigness syndrome, in conjunction with the felt sense that climate change is "too complicated," may lead one to conclude that climate change is too big to be solved. It is certainly too big for me to solve. Coupled with the thought that "I will be dead when all of this happens," it is easy to see how the bigness syndrome may discourage one from being seriously engaged with or concerned about the issue. The sense is that "There is no difference I can make and I will not be here to see what happens in any event." This broadens into a larger view of human inadequacy or a false humility that forgives individual and collective inaction. After all, "We are small, the world is big. How can we possibly matter in relationship to the climate?"

Of greater importance as an influence, and as one that cuts across many of the other influences we have discussed, the general cynicism problem impacts attitudes about many issues, not just climate change. It begins with the notion that "people in public roles are always telling me things that are not true." This generally coincides with a declining respect for public officials, public institutions, private institutions, and a near total mistrust for the media. Political scientists, for example, have suggested for some time that the media's heavy focus on the game of politics rather than on its substance starts a spiral of cynicism that directly causes an erosion of citizen interest, and ultimately, citizen participation. But bad times, policy failures, and corporate misdeeds throw plenty of gas on the fires ignited by the game coverage of politics. The failures of government and the private sector to address problems citizens are concerned about, the decline of the American middle class, a sluggish economy, partisan paralysis, and a general sense that things just are not getting better seems to persist in the United States today. This may feed a general public cynicism based on the perception of a declining capacity or lack of will within our government and in our political system to solve public problems. Such a perception, to the degree it is widely shared, fuels a rapid acceleration in the decline of public confidence in all institutions.

As our laundry list of influences on public opinion about climate change (it's complicated, temporal distortion phenomena, shortsightedness, the contrast effect, wishful thinking, confirmation bias, political partisanship, religious fundamentalism, the bigness syndrome, and cynicism) are considered in the context of the relative lack of public knowledge about or understanding of the science, it is perhaps amazing that the divide between the scientific consensus and public opinion is not in fact greater than it is. Remember, if you ask people if they believe global warming is happening, nearly 70 percent will say yes. If you ask them, based on the scientific evidence, if they believe the scientists have reached a consensus on the causes and effects of climate change, only a little over 50 percent will say yes. In some surveys, as many as six in ten think the scientists are still debating all of this. Even though the science is pretty well settled, it is obvious that it is not having nearly the impact on public perceptions as other influences. And of course, political partisanship or ideology is far more influential than the science with respect to the formation of public opinion. Emotions, values, beliefs, and a variety of distortions or misperceptions all matter more in shaping the public discourse than the work of climate scientists. This is important to understand and to address in a practical way if we hope to respond effectively to the climate crisis. As we said at the outset of this discussion about public opinion, a strategy to respond to climate change must include a reframing of the climate crisis as a national and global concern in such a way as to move the public perceptions to that place where they are no longer impediments to policy action but are informed sufficiently to enable progress. Based on our discussion, what might we be able to say about what such a strategy might include?

Where Responding to the Crisis Must Begin

Based on our discussion, and assuming that public support for and involvement in efforts to respond to the challenges of climate change is absolutely essential for success in either mitigating or adapting to it, how we communicate about climate change is important. In fact, we might say that communication appears to be one of the biggest obstacles to progress in managing the climate crisis in a nation divided along partisan lines and confused by a host of misperceptions. We might best begin, therefore, by knowing the audience. As we have seen, when it comes to climate change, or any issue for that matter, the American public is complex. There are multiple publics, not just one public, with multiple values, beliefs, and interests. A one-size-fits-all message about global warming may not be possible. Likewise, almost any single message or presentation of information that works with one segment of the public may turn others off entirely. But different messages designed for different publics may also be problematic in that they may lead to perceived inconsistency and confusion. So, how do we begin?

First, we might start with values and beliefs, because that is where most of our opinions begin. Scientists of course begin with the data and their findings. But truth be told, this is not a place where most of us begin to think about any issue. We begin with the bottom line. By bottom line, I mean some sense of what is of *value to me* here and now. One wants to know why I should care and why I should care right now. In other words, what real but practical evidence impacting my present life directly is there that something I value is at stake and I need to care about it right now? All of the scientific studies talking about all of the trends and all of the implications of them for the next two or three or four decades, no matter how dramatic and how persuasively presented, can be too subtle for too many. Scientists are readily able to explain the evidence and to project accurately its implications, but are often perplexed at why average citizens cannot follow them.

Some in the scientific community have, despite their reluctance to discuss anything but the scientific work they are doing, begun to perceive and acknowledge a need for scientists to convey what science actually knows for sure in a manner that is user friendly and more accessible to the general public. Their lack of efficacy in communicating to the general public about global warming has seemed at times to be both overwhelming and discouraging. But one is left to wonder, based on some of the preconditioned attitudes different publics have about science and scientists, if explaining the science more clearly would influence opinion much. Perhaps it is something other than, or more likely in addition to, the science that needs to be conveyed.

Social scientists perhaps can better understand the need to take the audience into consideration and to target the message to specific segments of the public. This is typically done in political ads and product advertisements. Identifying the values of the target audience and making appeals that align with those values makes perfect sense. One-size messaging does not fit all. For example, discussing climate change as an environmental issue or a scientific reality simply will not work with every audience. Remember our "six publics." Some are predisposed by partisanship or other influences to disregard science and to be suspicious of environmental concerns. The science will have little influence on the thinking of some religious fundamentalists, as we have seen. Thus, the challenge is to find things other than science or environmentalism that these audiences do care about, and in the present tense, and tie global warming to that.

One can talk about the science of climate change, its causes and consequences, and what must be done to address the issue. But that may not move the discussion forward as much as we would like. If you want people to care and to act, the suggestion here is we need to make the issue personally relevant to them. Connecting on values and experiences that bring us together—family, community, and nation—often opens up emotional and motivating bonds that humanize the messenger. Such a foundation will be necessary for a productive discussion about climate change and the challenges it presents. Talking in terms of concerns about the future of our children, the economic costs of inaction, the concept of responsibility to our families, our community, and our nation may tap into more fertile ground than a detailed discussion of the science may ever touch. This is especially true if the discussion is tightly focused on things we can see and experience today.

We can tie the things we are already seeing and experiencing into the conversation. Weather events like extreme droughts, floods, and intense storms and their experienced damages are real, and people care about them and see them in connection to their lives. The effects of extreme heat and/or extreme precipitation on agriculture (degrading soil, reduced harvests, the impact of disease and pests) are visible, and people do care. Climate extremes are already placing observable stress on water supplies and are placing our energy and physical infrastructure at risk. We can talk about threats, most already felt to some degree, to railroads, coastal ports, and tunnels, and the effects of extreme weather events on inland transportation systems. Human health (worsening asthma, allergies, and new types of infectious disease) are also becoming relevant in the present tense as related to a warming climate. The point is, with the importance and growing certainty of the science not to be ignored, the conversation that might actually be most helpful needs to focus on weather, agriculture, water, roads and bridges, human health, and other components of our experienced lives that are already being touched very directly by climate change. Making our lives and experiences the focus may open up a very meaningful dialogue that brings more people together and creates new partnerships to promote reasonable policy actions.

Perhaps the simplest advice for anyone to derive from our discussion with respect to public opinion and communication about climate change is that we must start with people and what they are experiencing. This means connecting what they care about to climate change and doing so in their own words, so to speak. Also, keep in mind that people understand what they see and experience. Discussions of a recently experienced tropical storm, a series of abnormally intense tornadoes, or an increasing number of wildfires are things people will understand almost immediately as important and timely. They've experienced it directly or seen it on the news. They get it. Don't start the global warming discussion at the planetary level; begin at the personal level. Again, around 70 percent of Americans believe the climate and the weather are changing. So begin not with science, for that will turn off those who are predisposed to oppose science for political or religious reasons. Begin with shared observations about the warmer summers, storm-filled winters, record droughts, intense wildfires, water shortages, and other impacts that people are feeling and experiencing right now.

Among those whom I believe may be ideally situated to start with, to begin at the personal level and build up to the planetary level, are individuals working in the field of emergency management. My general interests in that field of study certainly helped place climate change more squarely in front of me as a logical concern. With the sheer number of recurring and expected natural disasters and the costs they impose on communities across the nation, the impacts of climate change move front and center. The increasing frequency and the growing intensity of natural disasters and the rapidly changing risk and vulnerability profiles of every community across the nation in relation to a warming climate means there is much that is immediate and practical in the emergency manager's realm that of necessity must be discussed and analyzed. The mainstreaming of emergency management concerns into a discussion about climate change is a logical necessity. The examination of that issue from an emergency management perspective with a focus on assessing the real but changing risks and vulnerabilities to our communities, our businesses, our homes, its impacts on our economies, on our health, and the costs of disaster damages in general may attract far more attention, produce much more agreement, and create urgency about the need to act.

As a rule, and independent of ideology and politics, people think it is a good thing if their home or their business is not destroyed by a severe storm. They agree that it is good if a huge building or a bridge is not leveled by the wind or the shaking of the ground. Every community across the nation has an array of different but recurring natural disasters that they have experienced and that they expect to experience again. People living in these communities generally support being prepared for these events and think it a good idea to reduce risks and vulnerabilities they may impose. Now, as we shall see in some detail in Chapter 4, global warming has already had a role in changing the nature, the frequency, and the intensity of many of these disaster events we have experienced. It is expected that global warming will continue to influence the nature, frequency, and intensity of such events. In that context, a discussion of something everyone in a community is interested in or has had experience with, a natural disaster episode just experienced perhaps, provides an excellent opportunity to begin at the level of the personal and to tie it into the planetary level. People get it. They recognize what has just occurred. Many felt it directly and sustained losses. They know it will happen again. There always will be the next storm or the next flood. They will be interested in reducing their risks and their vulnerabilities. Connecting the warming climate as a factor in defining those risks and vulnerabilities, about which everyone is concerned, may be a means of bringing more people together in support of not only better hazard or disaster mitigation planning generally but also in support of mitigation planning with respect to a warming climate.

A recent survey commissioned by the Stanford Woods Institute for the Environment and the Center for Ocean Solutions found that an overwhelming majority of Americans want to prepare in order to minimize the damage likely to be caused by global sea level rise and storms. A majority also said they wanted people whose properties and businesses are located in hazard areas—not the government—to foot the bill for this preparation. More than eight out of ten Americans surveyed said that people and organizations should prepare for the damage likely to be caused by sea-level rise and storms, rather than simply deal with or pay for the damage after it happens. The most popular policy solutions identified in the survey were strengthening building codes for new structures along coastal areas to minimize damage (favored by 62 percent) and preventing new buildings from being built near the coast (supported by 51 percent). (34) It may be entirely possible that preparing for natural disasters (tropical storms, tornadoes, wildfires, flooding, winter storms, etc.) is an excellent place to begin a conversation that will tilt the topic of climate change and examine it through a new prism. An emergency management perspective, connected to the realities people are living in relation to natural hazard risks all over the country, may make the impacts of a warming climate more understandable to more people than any scientific or political communication. I am inclined to suggest that it may even produce enough of a consensus amongst the different publics across the nation to support some reasonable policy action in response to real risks and vulnerabilities understood by real people wherever they may reside, physically and on any political spectrum.

Communicating the risks and vulnerabilities to our communities and to our lives associated with a warming climate is the place to begin the dialogue about climate change. At least that is the conclusion I have reached in my own study and evaluation. As we have demonstrated through these first three chapters, we have been feeling the heat. The scientific research that has led to the scientific consensus about global warming is robust and indisputable. But we have also seen that the politics surrounding climate change and the things that influence public opinion have worked against implementing a policy agenda that is forward looking and that addresses the reality of climate change. In order to make progress on the policy front, it is at least in part necessary to reformulate or recast the reality of climate change so that it is indisputably real and important to the overwhelming majority of our citizens. What is to follow in the next chapter and beyond is an effort to do just that.

An emergency management perspective, the perspective that hooked me into this conversation, begins with an assessment of risks and vulnerabilities. Based on that assessment, actions may be taken to mitigate or reduce costly and tragic outcomes and to prepare for and respond to the expected impacts of natural (and human made) hazards. This is logical reasoning which, after our discussion of the science, the politics, and public opinion we may now be prepared to engage. Thus we will next turn our attention to risk and vulnerability assessment. This will be done from the personal, the community, and the planetary level, but with an emphasis on the personal, as that is the level at which any meaningful action must begin. We will then discuss in subsequent chapters the things we can and must do to mitigate, to prepare for, and to respond to the risks and vulnerabilities identified. Assuming that we are able to bring the personal and the planetary concerns closer together, we will conclude by discussing the broader policy implications and the prospects for progress at the general policy level.

Any progress we are able to make in responding to the warnings of the science ("Earth, we have a problem") will require that an overwhelming majority of our citizens favor taking preventive action. They must come to see that taking preventive and adaptive actions are not necessarily harmful to the economy or to any other practical concern that we may regard as important. As we noted earlier, progress depends on our awareness and our agreement that global warming or climate change is really happening, that this is a bad thing, that global warming is caused by humans, and that there are things we can do and need to do about it. The scientific research, our political leadership, and our public discourse has not taken us to the level that we can agree that we are at a critical point of decision which, if not handled in an appropriate and timely manner, or if not handled at all, may turn into a disaster or catastrophe of unprecedented proportions. Another type of conversation may be necessary or, at the very least, more helpful to communicate the crisis at hand and the practical need to act. Communicating the crisis at hand through this conversation may once and for all put an end to the false debate that continues to impede reasonable progress.

References

1. Revelle, R., et al. "Atmospheric Carbon Dioxide." In the President's Science Advisory Committee, Panel on Environmental Pollution, Restoring the Quality of Our Environment: Report of the Panel on Environmental Pollution. Washington, D.C.: The White House, 1965.

2. "Special Message to Congress on Conservation and Restoration of Natural Beauty." February 8, 1965, American Presidency Project. http://www.presidency.ucsb.edu/ws/index.php?pid=27285 (accessed July 10, 2013).

3. MacDonald, G., et al. (1979). "The Long Term Impact of Atmospheric Carbon Dioxide on Climate." Jason Technical Report JSR-78-07. Arlington, Va.: SRI International, I.

4. Charney, J., et al. (1979). "Carbon Dioxide and Climate: A Scientific Assessment." Report of an Ad Hoc Study Group on Carbon Dioxide and Climate, Woods Hole, Massachusetts, July 23–27, 1979, to the Climate Research Board, National Research Council. Washington, D.C.: National Academies Press.

5. Weart, S.R. (2008). *The Discovery of Global Warming*. Cambridge, Mass.: Harvard University Press.

6. Oreskes, N., and Conway, E. (2010). *Merchants of Doubt*. New York: Bloomsbury Press.

7. Jastrow, R., Nierenberg, W., and Seitz, F. (1991). *Scientific Perspectives on the Greenhouse Problem*. Ottawa, Illinois: Marshall Press.

8. Rahm, D. (2010). *Climate Change Policy in the United States*. Jefferson, North Carolina: McFarland and Company, Inc.

9. World Meteorological Association, World Climate Conference (1979). http://www.dgvn.de/fileadmin/user_upload/DOKUMENTE/WCC-3/Declaration_WCC1.pdf (accessed July 17, 2013).

10. Intergovernmental Panel on Climate Change (IPCC). http://www.ipcc.ch/ (accessed July 17, 2013).

11. Dressler A., and Parson, E.A. (2010). *The Science and the Politics of Global Climate Change*. Cambridge, UK: Cambridge University Press.

12. The Kyoto Protocol. http://unfccc.int/kyoto_protocol/items/2830.php (accessed July 17, 2013).

13. Global Warming Skeptics Make a Plan. http://www.euronet.nl/users/e_wesker/ew@shell/API-prop.html (accessed July 24, 2013).

14. Global Warming Petition Project. http://www.petitionproject.org/ (accessed July 24, 2013).

15. Rahm, D. (2010). *Climate Change Policy in the United States*. Jefferson, North Carolina and London: McFarland and Company, Inc.

16. U.S. Withdraws from Kyoto Protocol (2001). http://news.bbc.co.uk /2/hi/science/nature/1248278.stm (accessed July 17, 2013).

17. Frontline Climate of Doubt Timeline. http://www.pbs.org/wgbh /pages/frontline/environment/climate-of-doubt/timeline-the-politics-of -climate-change/ (accessed July 23, 2013).

18. Tillerson, R. (*New York Times* Interview, 2006). New Exxon-Mobil CEO Signals Change. http://www.nytimes.com/2006/03/30/business /worldbusiness/30iht-exxon.html?pagewanted=all&_r=0 (accessed July 31, 2013).

19. Exxon Cuts Ties to Global Warming Skeptics. http://www.nbcnews .com/id/16593606/ns/us_news-environment/t/exxon-cuts-ties-global-warming -skeptics/ (accessed August 24, 2015).

20. Spotts, P. (2013). "Climate change: Scientists now 95 percent certain we are mostly to blame." *The Christian Science Monitor* August 20, 2013. http://www.csmonitor.com/Environment/2013/0820/Climate-change -Scientists-now-95-percent-certain-we-are-mostly-to-blame (accessed August 27, 2013).

21. Intergovernmental Panel on Climate Change (IPCC). "U.N. Releases Definitive Report on Climate Change" November, 2007 http://www.ipcc.ch /publications_and_data/ar4/syr/en/contents.html (accessed August 27, 2013).

22. Vergano, D. (2013). "Climate Change: Stable Trends in Public Opinion Over Time." *USA Today*. http://www.usatoday.com/story/tech /columnist/vergano/2013/03/23/climate-polls-questions/2007471/ (accessed April 6, 2014).

23. Villar, A., and Krosnick, J.A. (2011). "Global Warming vs. Climate Change, Taxes vs. Prices." *Climate Change* 105:2–12.

24. Krosnick, J.A, and MacInnis, B. (2013). "Does the American Public Support Legislation to Reduce Greenhouse Gas Emissions?" Stanford University Public Opinion on Climate Change Project Report. http://climate publicopinion.stanford.edu/sample-page/research/does-the-american-public -support-legislation-to-reduce-greenhouse-gas-emissions/ (accessed April 6, 2014).

25. Roser-Renouf, C., Stenhouse, N., Rolfe-Redding, J., Maibach, E. and Leiserowitz, A. (2014). "Engaging Diverse Audiences with Climate Change: Message Strategies for Global Warming's Six Americas." In *Handbook of Environment and Communication*. Hanson, A., and Cox, R. (Eds.) London: Routledge.

26. International Panel on Climate Control (2014). "Climate Change 2014: Impacts, Adaptation, and Vulnerability: Summary For Policymakers." http://ipcc-wg2.gov/AR5/images/uploads/IPCC_WG2AR5_SPM _Approved.pdf (accessed June 27, 2014).

27. Yale Project on Climate Change Communication (2013). "Climate Change in the American Mind: Americans' Global Warming Beliefs and

Attitudes." http://climatechangecommunication.org/sites/default/files/reports
/Climate-Beliefs-April-2013.pdf (accessed June 28, 2014).

28. Pew Research Center Report (2013). "Deficit Reduction Rises on
Public's Agenda for Obama's Second Term: Public's Policy Priorities: 1994–
2013." http://www.people-press.org/2013/01/24/deficit-reduction-rises-on
-publics-agenda-for-obamas-second-term/#environment (accessed June 30,
2014).

29. Brulle, R.J., Carmichael, J., and Jenkins, J.C. (2012). "Shifting Public
Opinion on Climate Change: An Empirical Assessment of Factors Influencing
Concern over Climate Change in the U.S., 2002–2010." *Climatic Change*
(February).

30. The Weather Channel (2013). "Cold Snaps, Global Warming Go
Hand-in-Hand." http://www.weather.com/news/weather-winter/nasa-cold
-snaps-global-warming-20130129 (accessed June 30, 2014).

31. Coleman, P.T. (2011). "Climate Change, Partisanship and Conflict:
What's a Weather Beaten Nation to Do?" *Psychology Today* (October 30).

32. Freeman P. K., Houston D. J. (2011). "Rejecting Darwin and Support
for Science Funding." *Social Science Quarterly* 92, 1150–1168.

33. Stewart, K. (2012). "How the Religious Right Is Fueling Climate
Change Denial." *The Guardian* November 5. http://www.alternet.org
/environment/how-religious-right-fueling-climate-change-denial   (accessed
June 30, 2014).

34. Stanford Woods Institute for the Environment (2013) news release.
"Americans Back Preparation for Extreme Weather and Sea-Level Rise."
http://climatepublicopinion.stanford.edu/wp-content/uploads/2013/04
/Adaptation-News-Release.pdf (accessed July 1, 2014).

CHAPTER 4

Risks and Vulnerabilities: Making the Planetary Personal

Introduction

An emergency management perspective, as a general proposition, says that hazard mitigation or the ensuring of resilience in the face of predictable hazard risks and the promotion of economic, political, social, and environmental sustainability, is a necessary and a critical thing at the community level. This requires a full awareness of hazard risks and vulnerabilities and the formulation and implementation of plans to reduce them. The prevention of disaster damages, or at least the reduction of their costs to humanity and the environment, is an essential characteristic of a livable and sustainable community. Emergency managers know this. They also know that structural adjustments (e.g., building codes, structural engineering, retrofitting, etc.) to create disaster resilience and reduce the threat of damages may enable communities to withstand disaster impacts and recover from them more quickly. But these adjustments are not sufficient in and of themselves to promote human and environmental sustainability. It is also necessary to decide to refrain from or avoid activities that constitute direct threats to sustainability. It is necessary to promote human development and living strategies that preserve the socioecological system. This requires that human beings take responsibility for disasters. Identifying hazard risks and the potential for disaster resident in them is step one in taking responsibility. Thus, risk and vulnerability assessment is step one in the emergency management cycle. As indicated from the outset, this is the type of thinking that pulled me into the climate change discussion as a logical connection to my ongoing work in the field of emergency management.

On a very practical level, practicing emergency managers in every community across the nation can see the implications of climate change or a warming climate for the communities and the people they serve. A full awareness of hazard risks and vulnerabilities and the formulation and implementation of community-level plans to reduce them simply cannot be done without consideration of global warming. Increasingly, it is dawning on many that the effects of global warming might make more of an imprint on public thinking when they are considered locally as opposed to globally. The place where they are most likely to be considered locally is in relation to the differing natural hazards that contribute to recurring natural disasters.

The effects of global warming, those already felt and those that may be reasonably projected for the future, include the increased frequency and intensity of hurricanes, floods, wildfires, winter storms, heat waves, drought, food-borne and water-borne diseases, and the escalating of disruptions and disaster recovery costs to our local communities. These are real occurrences that real people are experiencing at ever more intense levels in their immediate lives. These are things that emergency managers must be prepared for in the normal course of their work. The normal course of their work, as it turns out, is increasingly connected to global warming.

The literature in the field of emergency management is discussing with greater frequency the linkage between global warming and new challenges it poses for risk assessment, risk management, disaster mitigation, disaster response, and disaster recovery. (1, 2, 3) *Climate-change preparedness*, by which is meant the assessing and preparing for climate-change impacts, is increasingly considered to be a part of disaster preparedness in general. (4, 5) Around the globe, practicing emergency management planners or experts and the communities they serve are being told they must confront the challenges presented to them by global warming as a practical and applied part of their work. (6) The assessment of its impact on local natural disasters and of its global impacts has been articulated as new necessities for emergency management professionals if they are to be successful in the normal course of their work. (7)

As emergency management professionals are beginning to seek out and incorporate into their normal operations scientific and predictive information relevant to maintaining the sustainability and resilience of the communities they serve in relation to the already felt and predictable future impacts of climate change, they are an instrumental part of translating the global challenge, the planetary climate crisis, into a local community problem to be solved. This is really best articulated perhaps as a personal concern for every local official, business, and home owner. The focus may be on drought, wildfires, severe weather, water shortages, flooding, and the variety of regularly recurring natural disaster or hazard potentials in each specific community, but each of these is a part of or connected to the planetary phenomenon of climate change. One need not worry about what the opinions of a local population are regarding the planetary crisis called climate change. It is enough to know and build on the fact that all residents in a community do care about the next flood or wildfire or storm, and their interest is more or less personal. They will have an intense interest in understanding local vulnerabilities and risks, in local disaster preparedness, and in mitigation or preventive measures that will reduce the vulnerabilities and risks in the communities where they work and live. Starting with the local, the next flood or wildfire or storm in a community may be the best place to begin a conversation that will eventually connect citizens to the planetary level and make climate change more immediate and real as a local and personal concern. In the same vein, a general discussion of the risks and vulnerabilities imposed on the planet and to humanity by global warming may have greater impact on public opinion and on policy makers if it personalizes them in a language that is understandable and a discussion of what people are actually experiencing. Hence, the greatest need may be for a discussion of climate-change risks and vulnerabilities from an emergency management perspective with a focus on people and the experiences that are both immediate and real to them.

The premise of this chapter is that understanding the planetary crisis represented by global warming requires, or is very significantly enhanced by, a connection to the lives and experiences of individuals and communities at a local level. Yes, we must assess and address risks and vulnerabilities at a global level, but we will understand them better and respond more intelligently as we localize them and personalize them. Thinking in terms of risk and vulnerability assessments, the first step in an emergency management perspective, we must find a way to make the planetary personal. Ultimately, this means in-depth assessments of the impact of climate change on each community and its risks, vulnerabilities, and felt impacts need to be localized. That is beyond the scope of this book, but it is actually practical work that is being done in communities all across the country. What we can do as we discuss the risks and vulnerabilities posed by climate change for the United States is articulate them from region to region across the nation with an emphasis on what they mean today and what they will imply tomorrow for the lives of people and the sustainability and resiliency of the communities across the country where they reside.

In our discussion of the risks and vulnerabilities associated with global warming, we shall do so with an emergency management focus on their implications for natural disasters and their human impact. We shall endeavor to discuss the planetary in personal terms as we proceed region by region across the country. One way to do that will be to discuss what the identified threats and vulnerabilities will entail for communities and emergency management practitioners and planners in various parts of the country. For them and the people they serve, it gets personal and local in a hurry.

Climate-Change Impacts and Projections

In May 2014, the U.S. Federal Government issued its newest national assessment of climate change. This federal climate assessment—the third since 2000—brought together hundreds of experts in academia and government to assess the scientific research with the goal of guiding U.S. policy based on the best available climate science. The authors of the more-than-800-page report said it aims to present "actionable science" and a road map for local leaders and average citizens to mitigate the effects of a warming climate. (8) This came on the heels of the fifth and most complete assessment of the global impacts of climate change by the International Panel on Climate Change (IPCC).

The IPCC fifth assessment reiterated that humans were the "dominant cause" of global warming and that to avoid catastrophic climate change, the world's fossil-fuel reserves would have to *stay in the ground*. More importantly, the impact of global warming on humans and on natural systems was emphasized as never before. Irreversible changes may already be in motion, according to the report. (9) For the first time, the IPCC has clearly articulated climate change as a series of predictable risks to humans and to natural systems that are escalating or multiplying with warming global temperatures. These risks were in turn discussed as they related to people. The gravest risks are to people in low-lying coastal areas and on small islands due to storm surges, coastal flooding, and sea level rise. Food supplies are also at increasing levels of risk due to drought, changing rainfall patterns, and flooding. Drought may also put safe drinking water in short supply, and more severe storms will threaten electricity stations and damage infrastructure. (9) Changes in climate have already had measurable negative impacts on humans and natural systems on all continents according to the report, and more dramatic impacts are expected. As expected, many Americans yawned at yet another IPCC report, and conservative contrarians quickly labeled it as alarmist nonsense. But the U.S. climate report that was released a couple of months after the IPCC report brought the impacts of climate change into an even more immediate and personal perspective for many Americans.

The U.S. climate report of 2014 emphasized that global warming is affecting every part of the United States right now. This emphatically dispelled the notion that our concerns about a warming climate are far off or of significance only to polar bears and to our great-grandchildren. The report noted the general impacts: increased sea level rise, stronger storm surges, more severe flooding, more intense and frequent heat waves, changes in precipitation patterns, impact on water supply, hurricanes in the Southeastern United States, and more drought and wildfires in the Southwest. But it did so with a more detailed focus on local conditions already experienced across the country and targeted projections for the future. Before we examine these details from a localized perspective and begin to think specifically in terms of risk and vulnerability assessments, let us take note of the broad strokes in the general assessment that are easier to ignore than the immediate conditions we may experience in any specific location at any given time.

The U.S. climate report, to be sure, echoes the findings of the report issued by the IPCC and its scientists. U.S. scientists have concurred with scientists globally that the warming experienced over the past 50 years has been caused mostly by emissions of heat-trapping gasses by humans. By the end of the century, the U.S. report says, temperatures can increase by 5 degrees. Even if we were to act aggressively to reduce greenhouse gas emissions, temperatures could increase by as much as 10 degrees if emissions continue unabated. Extreme weather caused by global warming has increased in recent decades and will be expected to continue to do so, according to the report. (8) More important than these general conclusions, at least in the context of our discussion in this chapter, is the assessment of impacts more locally across the United States.

The U.S. report, for just one example, articulated major concerns for people living in the mid-Atlantic region. Among the specific things discussed for that region was the expected increase in coastal flooding and the destruction of wetlands that protect against storm surge. Given increasing greenhouse gas emissions, the majority of Maryland and Delaware, and parts of West Virginia and New Jersey, are projected to have 60 additional days of 90-degree temperatures by the middle of the century. (8) The report, as we shall see, has specific assessments for each region of the country. This means that the impacts, which will vary from one part of the nation to the next, will be very real to people in every community across the country. That conversation—what it means to me where I live—combined with the experiences we are already having as the climate changes before our eyes with each natural disaster that unfolds in more dramatic fashion in the normal course of events, may be the key to inspiring public concern and policy action.

Unfortunately and predictably, the release of the 2014 national climate report generated the expected response from conservative contrarians. They called the report biased, suggested it was a political document designed to justify federal regulation of greenhouse gas emissions, and a part of a political agenda by liberals to implement an unchecked regulatory agenda that would cost the nation jobs and defeat its chances of achieving energy independence as it destroyed the economy. This is the standard contrarian response to each new report about the predicted impacts of climate change. The matter might well have been left there as another episode in the battle between science and everything not related to science that some ideologues can dream up to throw against it. But this time there may be a critical difference that has the potential to change the dialogue. This time the planetary might be more personal, and the science may be aided by the actual experiences of real people who are actually feeling the heat right where they live. Average citizens need much more than science they don't quite understand or politicians they neither like nor trust to help them get a handle on global warming. They need a practical assessment of the risks and vulnerabilities relevant to them and where they live. That, it is suggested here, may successfully compete for their serious attention. It would most certainly help citizens to see the predicted impacts of climate change as related to them and their immediate lives.

U.S. Risks and Vulnerabilities by Region

Let us turn now to a summary of the 2014 U.S. Climate Report. As we discuss the risks and vulnerabilities associated with global warming for each region of the United States, we will begin with a brief summary excerpt for each region from the report. This excerpt for each region will be followed by an elaboration or a summary of the general conclusions reached by the scientists, who have carefully reviewed and presented the scientific evidence that supports them, in terms of the sorts of concerns an emergency management perspective must include. This, as we have suggested, may be of great value in efforts to make the global or planetary issue of global warming a personal issue for more and more Americans.

Northeast

1. Climate Risks to People
   Heat waves, coastal flooding, and river flooding will pose a growing challenge to the region's environmental, social, and economic systems. This will increase the vulnerability of the region's residents, especially its most disadvantaged populations.
2. Climate Risks to Infrastructure
   Infrastructure will be increasingly compromised by climate-related hazards, including sea level rise, coastal flooding, and intense precipitation events.
3. Climate Risks to Agriculture and Ecosystems
   Agriculture, fisheries, and ecosystems will be increasingly compromised. Farmers can explore new options, but these adaptations are not cost- or risk-free. Moreover, adaptive capacity, which varies throughout the region, could be overwhelmed by climate change.
4. Planning and Adaptation
   While a majority of states and a rapidly growing number of municipalities have begun to incorporate the risk of climate change into their planning activities, implementation of adaptation measures is still at early stages. (2014 U.S. Climate Report)

Sixty-four million people live in the American Northeast. The urban corridor from Washington, D.C., to Boston is highly developed and contains a massive and complex network of supporting infrastructure. Two recent events in this region of the country have given residents fresh experiences with vulnerability to extreme weather-related events. Hurricane Irene, in August 2011, produced a broad swath of very heavy rain, greater than five inches in total and sometimes two to three inches per hour in some locations, from southern Maryland to northern Vermont. Hurricane Sandy caused massive coastal damage from storm surge and flooding along the Northeast coast in October 2012. During Irene, the New York City transit system was shut down, and over 2.3 million coastal residents in Delaware, New York, and New Jersey were evacuated. Inland impacts were also severe. Flash flooding washed out roads and bridges, flooded homes and businesses, brought down power lines, damaged barns, and flooded crop fields. Sandy caused over $65 billion in damages to the region. These damages included 650,000 homes damaged or destroyed, 8.5 million people without power, the inundation of New York City subway tunnels with floodwaters, and significant damage to the electrical grid and sewage treatment plants. (8) These recent experiences, fresh in the memory of residents, can be expected to recur and with growing intensity as a result of global warming. Local planners and leaders can use that to good advantage with respect to building public support for preparedness and mitigation measures that a warming climate will make necessary.

The scientists are convinced that heat waves, heavy downpours, more severe winter storms, and sea level rise pose predictable and growing challenges to all aspects of life in the Northeast. More serious disaster-related impacts are expected, and infrastructure, agriculture, fisheries, and ecosystems will be increasingly compromised. Considering this, along with the recent experiences of the region, it is necessary from an emergency management perspective to factor these climate-induced risks into local planning for future disasters. Both with respect to preparing for a disaster occurrence and with respect to taking steps to reduce or mitigate the damage of future events, climate change must be a practical component and a consideration in all local discussion and planning. The discussion can emphasize some simple but basic points reinforced by recent experience. Climate change is not a far-off concern. It is not a matter of scientific uncertainty. It is not a political thing. It is happening before our eyes. It is changing weather patterns. We have already been shown by our recent natural disaster experiences that we are vulnerable, and it is a certainty that we will experience another natural disaster. Including the predictable impacts of a warming climate on our communities in relation to the projection of future disaster impacts is a logical necessity.

States in the Northeast have already begun the discussion in earnest. Of the twelve states in the Northeast, eleven are developing climate-change adaptation plans for specific sectors, and ten are planning to release statewide adaptation plans. Northeast cities have begun to employ a variety of mechanisms to respond to climate change. These include land-use planning, provisions to strengthen and protect infrastructure, regulations related to the design and construction of buildings, and emergency preparation, response, and recovery planning that incorporates the risks associated with climate change. (8) All of this is an essential part of localizing the planetary, and it helps to make it very immediate and personal for residents.

The state of Vermont, as an interesting example, has been working on a project to prepare a comprehensive group of recommendations to increase Vermont's preparedness for the effects of climate change and its extreme weather impacts. These recommendations, contained in a document entitled *Vermont's Roadmap to Resilience*, are meant to prepare for natural disasters and the effects of climate change. The stated purpose is to identify practical steps to reduce vulnerabilities and minimize the risks to citizens, communities, the economy, and the environment. (10) Government officials, business leaders, community stakeholders, and concerned citizens have all participated in the Resilient Vermont Project. The Resilient Vermont Project was initiated largely in response to extreme weather events in 2011—the unprecedented spring flooding and Tropical Storm Irene in August. This project was a collaborative effort between the Institute for Sustainable Communities (ISC), the State of Vermont, and several other partners to develop priority recommendations for next steps in resilience in relation to climate change. Over an 18-month period, risks and vulnerabilities were assessed and specific recommendations were developed. Interestingly, these recommendations included the elevation and integration of emergency management into all local, regional, and statewide activity related to planning for resilience and responding to the risks of a warming climate. (10) The focus of the effort was to proactively reduce vulnerabilities, improve disaster response and recovery, and strengthen resilience in the face of natural disasters and climate-related shocks and the identifiable risks they impose on the people and communities of Vermont. (10) Vermont's Roadmap to Resilience is an excellent example of making the planetary local. Every state and every community in the nation will, as we shall see, benefit from undertaking a similar effort.

Southeast

1. Sea Level Rise Threats
   Sea level rise poses widespread and continuing threats to both natural and built environments and to the regional economy.
2. Increasing Temperatures
   Increasing temperatures and the associated increase in frequency, intensity, and duration of extreme heat events will affect public health, natural and built environments, energy, agriculture, and forestry.
3. Water Availability
   Decreased water availability, exacerbated by population growth and land-use change, will continue to increase competition for water and affect the region's economy and unique ecosystems. (2014 U.S. Climate Report)

Over 70 million people live in the American Southeast. The Southeast includes 29,000 miles of coastline and many cities. Jacksonville, Charlotte, Atlanta, Miami, and New Orleans have populations of over 250,000. Densely populated coastal areas and coastal ecosystems in the Southeast are already experiencing relative sea level rise, hurricanes, and worsening storm surge. Climate change is projected to exacerbate these existing threats significantly. The region's economy, which includes forestry, tourism, oil and gas production, and agriculture, will be impacted by climate change as well.

Portions of the Southeast and Caribbean are extremely vulnerable to sea level rise. This is because a large numbers of cities, roads, railways, ports, airports, oil and gas facilities, and water supplies are at low elevations and potentially vulnerable to the impacts of sea level rise. Half of the residents of New Orleans, for example, live below sea level. As Hurricane Katrina demonstrated in 2005, this makes the city extremely vulnerable to the impacts of tropical events. Miami, Tampa, Charleston, and Virginia Beach are among U.S. cities considered to be most at risk in relation to rising sea levels and/or the increasingly more devastating impacts of storm surges. (8) It must also be noted that sea level rise and storm surge can have impacts far beyond the area directly affected. Homes and infrastructure in low-lying areas are also increasingly prone to flooding during tropical storms.

There are other factors that increase the risks associated with rising sea levels, such as wetland loss, and these seriously compromise the protection of people and infrastructure in coastal areas. And people and infrastructures are already plenty compromised. It is estimated that coastal counties and parishes in Alabama, Mississippi, Louisiana, and Texas, with a population of approximately 12 million, assets of about $2 trillion, and producers of $634 billion in annual gross domestic product, already face significant losses in relation to sea level rise. These current losses average about $14 billion annually. By 2030, assuming no increase in hurricane intensity or additional sea level rise, these losses could reach $18 billion per year. With a mere three percent increase in hurricane wind speed and just under six inches of sea level rise, these losses will exceed $23 billion annually. (8) More frequent storm surge flooding of low-lying areas can be expected to cause more frequent flooding of the transportation infrastructure throughout the region. This can disrupt travel and cause increasing damages to roads, highways, bridges, and oil and gas operations in coastal areas. The transportation network is especially vulnerable because many roads in the coastal areas are at a low elevation, four feet or less, and subject to frequent flooding. These risks and vulnerabilities have all existed in this region, and the people living there have experienced them first hand. The point is, climate change will greatly enhance them. As a matter of local concern, that is a reality that cannot be escaped. (8)

The impact of rising temperatures will be acutely felt in the Southeast. From a public health perspective, Atlanta, Miami, New Orleans, and Tampa have already had increases in the number of days with temperatures exceeding 95 degrees F. This has correlated with an increase in the number of heat related deaths. (8) A significant rise in hospital admissions due to respiratory illnesses, emergency room visits for asthma, and lost school and work days are among the expected by-products in this region of a warming climate. As the number of hot days increase in frequency and intensity, the occurrence and duration of extreme heat events will affect public health. Summer heat stress will reduce crop productivity, lead to more severe drought conditions, and pose threats to fresh water supply. (8)

As each community in the Southeast begins to assess the specific risks it faces with respect to climate change, they will find that they are examining risks they have always known about in the normal course of events. They will see that some of what they are already experiencing in relation to natural hazards has already been impacted by a warming climate. They will also, as they anticipate and examine the projected or future impacts of a warming climate, see that climate change is taking what they have previously experienced and what they know to expect for the future (the next hurricane, the next flood, the next drought) to new and more devastating levels. Individuals and communities throughout the Southeast are already discussing strategies to adapt to, accommodate, or protect against the impacts of a rising sea level. These discussions of local conditions and the need to adapt to them, protect the local infrastructure, and promote the resiliency of our communities and local economies are where global warming is becoming a very important local concern. People will generally not object to acting proactively to reduce vulnerabilities, improve response and recovery, and strengthen resilience in relation to natural disasters.

As it becomes increasingly clear that sea level rise poses widespread and continuing threats to the region's economy and environment; that extreme heat will affect public health, energy, agriculture, and more; and as the felt impact of recurring disaster events common to the region but newly intensified by the influences of a warming climate increase, it will be abundantly clear that in the assessment of local concerns we have already been looking at the planetary and making it personal. Local planners and emergency management professionals will need to anticipate the impacts of a warming climate, enhance or improve threatened infrastructure, make adjustments in land use and agricultural practices, protect the water supply, manage public health challenges, be prepared to respond efficiently to recurring and more frequent disaster events, and engage their communities in a dialogue that is aimed at reducing vulnerabilities. Whatever one's view about climate change, doing all of these things is always necessary, and they are expected within communities across the nation. Incorporating climate-related risks and vulnerabilities into these already existing concerns and responsibilities should not require much imagination. Indeed, doing so should easily come to be seen as an urgent necessity.

Midwest

1. Impacts to Agriculture
   In the next few decades, longer growing seasons and rising carbon dioxide levels will increase yields of some crops, though those benefits will be progressively offset by extreme weather events. Though adaptation options can reduce some of the detrimental effects, in the long term, the combined stresses associated with climate change are expected to decrease agricultural productivity.

2. Forest Composition
   The composition of the region's forests is expected to change as rising temperatures drive habitats for many tree species northward. The role of the region's forests as a net absorber of carbon is at risk from disruptions to forest ecosystems, in part due to climate change.
3. Public Health Risks
   Increased heat wave intensity and frequency, increased humidity, degraded air quality, and reduced water quality will increase public health risks.
4. Fossil-Fuel-Dependent Electricity System
   The Midwest has a highly energy-intensive economy with per capita emissions of greenhouse gases more than 20 percent higher than the national average. The region also has a large and increasingly utilized potential to reduce emissions that cause climate change.
5. Increased Rainfall and Flooding
   Extreme rainfall events and flooding have increased during the last century, and these trends are expected to continue, causing erosion, declining water quality, and negative impacts on transportation, agriculture, human health, and infrastructure.
6. Increased Risks to the Great Lakes
   Climate change will exacerbate a range of risks to the Great Lakes, including changes in the range and distribution of certain fish species, increased invasive species and harmful blooms of algae, and declining beach health. Ice cover declines will lengthen the commercial navigation season. (2014 U.S. Climate Report)

More than 61 million people live in the American Midwest. Home to expansive and productive agricultural lands, forests, the Great Lakes, substantial industrial activity, and major urban areas, the region generates a GNP that is about 19 percent of the national total ($2.6 trillion). (8) As is the case in other regions, climate change will tend primarily to amplify existing climate-related risks to people and infrastructure throughout the Midwest. Several types of extreme weather events have already increased in frequency and/or intensity due to climate change, and further increases are projected. Heat stress, more intense flooding, late spring freezes, and a variety of extreme weather events are already contributing to eco-system disturbance and infrastructure damages. An increase in atmospheric pollutants and economic shocks such as crop failure or reduced yields are becoming more common. More extreme weather events are also contributing to an escalation of disaster-related costs being borne by residents and communities. As in other regions, the infrastructure is negatively impacted. Most of the region's population lives in cities that are especially vulnerable to climate-change-related flooding and heat waves because of an aging infrastructure. Also, the region's energy-intensive economy emits a disproportionately large amount of greenhouse gases that contribute to the warming of the climate. (8)

In anticipation of more extreme weather (rainfall, flooding, storms), more intense heat waves, and the associated risks and vulnerabilities these climate-related events may mean for public health, infrastructure damage, agricultural production, and the resilience of its communities and its ecosystems, the Midwest will need to incorporate climate change into its normal planning and preparedness for recurring natural disaster events. Efficient response to recurring disaster events will require here, as in each region of the country, that planners engage their communities in a dialogue that is aimed at reducing vulnerabilities. That is, as we have repeatedly said, an ongoing or normal activity in any community. But it is an activity that must, if it is to be done successfully, now include in its normal course the accelerating impacts of a warming climate. Reduced crop yields, threatened infrastructure, degraded air and water quality, storm damages, public health risks, flooding, and the disruption of ecosystems are the things Midwestern communities are already experiencing that make the planetary local and very real to them.

The Great Plains

1. Energy, Water, and Land Use
   Rising temperatures are leading to increased demand for water and energy. In parts of the region, this will constrain development, stress natural resources, and increase competition for water among communities, agriculture, energy production, and ecological needs.
2. Agriculture
   Changes to crop growth cycles due to warming winters and alterations in the timing and magnitude of rainfall events have already been observed; as these trends continue, they will require new agriculture and livestock management practices.

3. Conservation and Adaptation
   Landscape fragmentation is increasing, for example, in the context of energy development activities in the northern Great
   Plains. A highly fragmented landscape will hinder adaptation of
   species when climate change alters habitat composition and
   timing of plant development cycles.
4. Vulnerable Communities
   Communities that are already the most vulnerable to weather
   and climate extremes will be stressed even further by more frequent extreme events occurring within an already highly variable
   climate system.
5. Opportunities to Build Resilience
   The magnitude of expected changes will exceed those experienced in the last century. Existing adaptation and planning
   efforts are inadequate to respond to these projected impacts.
   (2014 U.S. Climate Report)

The Great Plains is a region where water is a constant and critical
concern. It is a significant variable in every aspect of local life. Changes
in the weather can be dramatic and challenging in this region for communities and their commerce. Changes in weather can impact the region's water resources very dramatically. The Great Plains experiences
a variety of climate and weather hazards in its normal course of natural events. These include floods, droughts, severe storms, tornadoes,
hurricanes, and winter storms. In much of the region, too little precipitation falls to replace that needed by humans, plants, and animals.
These variable conditions in the Great Plains already stress communities and cause billions of dollars in damage. Climate change will add to
both stress and costs as it alters and makes more intense or severe the
impact of these recurring events. (8)

Great Plains residents and communities deal with a wide variety of
weather extremes on a recurring basis. They face constant challenges
from winter storms, extreme heat and cold, severe thunderstorms,
drought, and flood-producing rainfall. This is all a part of the normal
experience of the region. Climate change is already making that experience more dramatic and more costly throughout the region.
Projected changes in weather patterns associated with climate change
include significant increases in winter and spring precipitation in the
Northern plains states and longer dry spells and more severe droughts
in Southern areas such as Texas and Oklahoma. Projected increases
in extreme heat will have many negative consequences, including

increases in surface water losses, heat stress, and demand for air conditioning. There may be some positive winter effects such as warmer winters, lower winter heating demand, less cold stress on humans and animals, and a longer growing season. But overall rising temperatures will lead to increased demand for water. Changing extremes in precipitation are projected across all seasons, including higher likelihoods of both increasing heavy rain and snow events in some locations and more intense droughts in others. Communities vulnerable to extreme weather will experience more frequent and dramatic events within an already variable climate system and an increase in damage-related costs to infrastructure.

The residents and communities of the Great Plains face increasing climate-related vulnerabilities. Expected changes will exceed previous experiences with climate-related hazards and disasters. Water systems already under stress will be stressed all the more, agriculture will endure changes in crop growth cycles due to warming winters and changing precipitation patterns, and community resources and resilience will be more severely tested by the increasing frequency and intensity of floods, droughts, and heat waves. These projected impacts must be integrated into all local and community planning and development. As an emergency management proposition, this localization of the planetary change in climate must see Great Plains communities proactively engaged in a dialogue that is aimed at reducing vulnerabilities and promoting the resilience of the region.

Southwest

1. Reduced Snowpack and Stream Flows
   Snowpack and stream flow amounts are projected to decline in parts of the Southwest, decreasing surface water supply reliability for cities, agriculture, and ecosystems.
2. Threats to Agriculture
   The Southwest produces more than half of the nation's high-value specialty crops, which are irrigation-dependent and particularly vulnerable to extremes of moisture, cold, and heat. Reduced yields from increasing temperatures and increasing competition for scarce water supplies will displace jobs in some rural communities.
3. Increased Wildfire
   Increased warming, drought, and insect outbreaks, all caused by or linked to climate change, have increased wildfires and

impacts to people and ecosystems in the Southwest. Fire models project more wildfire and increased risks to communities across extensive areas.

4. Sea Level Rise and Coastal Damage

   Flooding and erosion in coastal areas are already occurring, even at existing sea levels, and damaging some California coastal areas during storms and extreme high tides. Sea level rise is projected to increase as Earth continues to warm, resulting in major damage as wind-driven waves ride upon higher seas and reach farther inland.

5. Heat Threats to Health

   Projected regional temperature increases, combined with the way cities amplify heat, will pose increased threats and costs to public health in southwestern cities, which are home to more than 90 percent of the region's population. Disruptions to urban electricity and water supplies will exacerbate these health problems. (2014 U.S. Climate Report)

The Southwest, with a population of 56 million people, is the hottest and driest part of the United States. Climate change is already impacting this parched region, and it is a certainty that it will get hotter and significantly drier in its southernmost part. Severe and sustained drought will stress water sources beyond anything previously known. This will increase competition among farmers, energy producers, urban dwellers, and plant and animal life for the region's most precious resource. (8) The impacts of drought conditions will have dramatic economic consequences as reduced crop yields from increasing temperatures and increased competition for water in rural communities will displace jobs in rural communities. The already experienced and the projected increase in wildfire outbreaks will bring increased damages and recovery costs to people and communities, and the cost of fire suppression will rapidly accelerate as well. Projected increases in extreme flooding in coastal areas (e.g., Southern California) as a result of sea level rise will increase human vulnerability to coastal flooding events. Projections in sea level rise make it abundantly clear that vulnerable populations and endangered infrastructure in coastal areas are increasingly at risk.

As if to accentuate the conclusions reached in the U.S. Climate Report about this region, the spring of 2014 saw one of the worst droughts ever experienced in the region, impacting California, Texas, Arizona, Oklahoma, and New Mexico. Two other states (Kansas and

Nevada) also experienced record drought. Drought conditions were so severe that each state had more than 50 percent of its land area in severe drought. Severe drought is characterized by heavy crop loss, frequent water shortages, and mandatory water use restrictions. It is entirely possible that much of the American Southwest is facing a future of chronic and perhaps even perpetual drought.

Over 90 percent of the people living in the Southwest reside in cities. This represents the highest urban population rate in the United States. Large metropolitan populations already pose challenges to providing adequate domestic water supplies. The combination of projected population growth and projected increased risks to surface water supplies posed by a warming climate will add further challenges. Conserving water to help meet the demands of a growing population and providing adequate water for urban greenery to reduce increasing urban temperatures will be competing priorities. Infrastructure concerns include the probability that extensive use of air conditioning to deal with high temperatures will quickly increase electricity demand and trigger cascading energy system failures, resulting in blackouts or brownouts. (8)

According to the U.S. Climate Report, heat stress (a recurrent health problem for urban residents) has been the leading weather-related cause of death in the United States since 1986. The highest rates of heat-related deaths nationally are found in Arizona. Exposure to excessive heat can also aggravate existing human health conditions. For those who suffer from respiratory or heart disease, exposure to excessive heat can be deadly. Increased temperatures can reduce air quality because atmospheric chemical reactions proceed faster in warmer conditions. The outcome is that heat waves are often accompanied by increased ground-level ozone, which in turn can cause respiratory distress. (8)

Like every other region of the country, many of the risks and vulnerabilities people and communities in the Southwest must deal with in the normal course of their climate variations are enhanced or made much worse by climate change. The already felt increase in drought conditions, the impact of more widespread and intense wildfires, the health, agricultural, economic, and infrastructure vulnerabilities already known, and those projected for the future may suggest that the American Southwest is ground zero for some of the most intense and immediate threats associated with a warming climate. The local manifestations of the planetary are compelling for the people and communities of this region. As their local planners and emergency management professionals assess the risks and vulnerabilities of this region, they cannot begin to do an adequate job without factoring into their calculations the impacts of climate change.

Northwest

1. Water-related Challenges
   Changes in the timing of stream flow related to changing snow-melt have been observed and will continue, reducing the supply of water for many competing demands and causing far-reaching ecological and socioeconomic consequences.
2. Coastal Vulnerabilities
   In the coastal zone, the effects of sea level rise, erosion, inundation, threats to infrastructure and habitat, and increasing ocean acidity collectively pose a major threat to the region.
3. Impacts on Forests
   The combined impacts of increasing wildfire, insect outbreaks, and tree diseases are already causing widespread tree die-off and are virtually certain to cause additional forest mortality by the 2040s and long-term transformation of forest landscapes. Under higher emissions scenarios, extensive conversion of sub-alpine forests to other forest types is projected by the 2080s.
4. Adapting Agriculture
   While the agriculture sector's technical ability to adapt to changing conditions can offset some adverse impacts of a changing climate, there remain critical concerns for agriculture with respect to costs of adaptation, development of more climate-resilient technologies and management, and availability and timing of water. (2014 U.S. Climate Report)

The Northwest's varied topography includes coastal areas, volcanic mountains, and high sage deserts. As is the case with each region of the country, the Northwest's economy, infrastructure, natural systems, public health, and vitally important agriculture sector all face important climate-change-related risks. Given the nature of the region and the complexity of its climate and geographic, social, and ecological features, impacts will vary significantly across the region. For example, impacts on infrastructure, natural systems, human health, and economic sectors, combined with issues of social and ecological vulnerability, will play out quite differently in largely natural areas like the Cascade Range or Crater Lake National Park than in urban areas like Seattle and Portland.

Warming in the Northwest has been linked to changes in the timing and amount of water availability in basins. Higher temperatures are causing more winter precipitation to fall as rain rather than snow, and this contributes to the earlier melting of snow. Further declines in snowpack are projected. This is critically important, as it reduces the amount of water available during the summer or warm season. This means that competition for water is expected to intensify, including for use by and for municipalities, industry, agriculture, hydropower production, navigation, recreation, and in-stream flows. The largest water-related impacts are expected to continue to occur in basins with significant snow accumulation. Warming increases winter flows and advances the timing of spring melt, which, in turn, contributes to imbalances in water supply because of reductions in summer flow. This presents a host of problems related to agriculture and ecology. (8)

Regional power planners have also expressed concerns over the existing hydroelectric system's potential inability to provide adequate summer electricity given the combination of climate-change impacts on water and the growth in demand. Much of the electricity in the Northwest is supplied by hydroelectricity. Decreasing summer stream flows will continue to reduce hydroelectric supply and stress electricity supplies. Meanwhile, rising temperatures will increase electricity demand for air conditioning and refrigeration. (8)

Northwest agriculture is very sensitive to climate change because of its dependence on irrigation water; a specific range of temperatures, precipitation, and growing seasons; and the sensitivity of crops to temperature extremes. Projected warming will reduce the availability of irrigation water in snowmelt-fed basins and increase the probability of heat stress to field crops and fruit trees. Higher average temperatures (and these are projected to increase) will exacerbate the number and variety of threats and increase the vulnerabilities of plant life to pest infestation, disease, and weed complexes. (8)

Global warming is associated with changes in flood risks. Changing precipitation patterns (regional models project increases in expected rainfall), increased winter flows, and earlier snow melts will all contribute to increased flood risk and an acceleration of accompanying flood-related damages and costs. The coastal zone will experience the effects of sea level rise, erosion, inundation, and the accompanying threats to infrastructure. Sea level rise will increase erosion of the coast and cause the loss of beaches and significant coastal land areas. Among the most vulnerable parts of the coast is the heavily populated south Puget Sound region of Washington, which includes the cities of Olympia, Tacoma, and Seattle. (8)

Warmer temperatures and less summer soil moisture will increase the likelihood of large, intense wildfires in some areas and reduce the ability of some species to regenerate because of changing habitats. From a human perspective, the costs associated with climate-change-induced wildfires will escalate significantly. Wildfires have been a perennial problem in the Northwest. With miles and miles of wilderness, Idaho, Montana, Wyoming, Oregon, and Washington all experience more than their fair share of major forest fires in the normal course of events. The impact of global warming will only increase the risks and vulnerabilities associated with wildfires as well as their frequency and severity. (8)

Like every other region of the country, the American Northwest is already experiencing the impacts of a warming climate. The local experience is making the planetary a personal concern for residents, and it is impacting their communities very directly. As these impacts continue to intensify, it will be imperative from what we are calling an emergency management perspective to be more acutely aware of the changing risks and vulnerabilities and to take steps to reduce them with wise planning and development strategies to promote resilience in the face of the threats associated with climate change.

Alaska

1. Disappearing Sea Ice
   Arctic summer sea ice is receding faster than previously projected and is expected to virtually disappear before mid-century. This is altering marine ecosystems and leading to greater ship access, offshore development opportunity, and increased community vulnerability to coastal erosion.
2. Shrinking Glaciers
   Most glaciers in Alaska and British Columbia are shrinking substantially. This trend is expected to continue and has implications for hydropower production, ocean circulation patterns, fisheries, and global sea level rise.
3. Thawing Permafrost
   Permafrost temperatures in Alaska are rising, a thawing trend that is expected to continue, causing multiple vulnerabilities through drier landscapes, more wildfire, altered wildlife habitat,

increased cost of maintaining infrastructure, and the release of heat-trapping gases that increase climate warming.

4. Changing Ocean Temperatures and Chemistry
   Current and projected increases in Alaska's ocean temperatures and changes in ocean chemistry are expected to alter the distribution and productivity of Alaska's marine fisheries, which lead the United States in commercial value.

5. Native Communities
   The cumulative effects of climate change in Alaska strongly affect Native communities, which are highly vulnerable to these rapid changes but have a deep cultural history of adapting to change. (2014 U.S. Climate Report)

Alaska has actually warmed twice as fast as the rest of the nation. This has had widespread and fairly dramatic impacts. Sea ice is receding rapidly, and glaciers are shrinking ever more quickly. Thawing permafrost is leading to more wildfire and affecting infrastructure and wildlife habitat. Rising ocean temperatures and acidification will alter valuable marine fisheries. (8) Since 1976, Alaska has experienced a fairly rapid warming, and climate-change impacts on Alaska are already pronounced. These include earlier spring snowmelt, reduced sea ice, widespread glacier retreat, warmer permafrost, and drier landscapes.

Some people might argue that there are positive impacts associated with climate change in Alaska. With reduced ice extent, the Arctic Ocean is more accessible for marine traffic, including trans-Arctic shipping, oil and gas exploration, and tourism. This will facilitate access to the substantial deposits of oil and natural gas under the seafloor in the Beaufort and Chukchi seas. But this "positive" would be accompanied by some negatives, such as raising the risk to people and ecosystems from oil spills and other drilling and maritime-related accidents. The biggest negative would be perhaps the acceleration of fossil-fuel extraction from the Arctic at a time when we need to leave fossil fuels in the ground and make a transition to renewables. A seasonally ice-free Arctic Ocean will also increase sovereignty and security concerns as a result of potential new international disputes and increased possibilities for marine traffic between the Pacific and Atlantic Oceans. (8)

The global decline in glacial and ice-sheet volume is projected to be one of the largest contributors to global sea level rise during the next century. Even if the air temperatures were to remain at current levels, glaciers continue to respond to climate warming that has already been experienced for decades after warming ceases. This means ice loss is expected to continue even if air temperatures were to remain at current levels or even cool a bit. With respect to Alaska, glacial retreat is currently increasing river discharge and hydropower potential in the south and central parts of the state. Over the long haul, however, glacial melting may well reduce water input into reservoirs and reduce hydropower resources. (8)

Changes in terrestrial ecosystems in Alaska and the Arctic may be negatively influencing the global climate system. Permafrost soils throughout the entire Arctic contain almost twice as much carbon as the atmosphere. This means that as the warming and thawing of these soils increases, the release of carbon dioxide and even more deadly methane through increased decomposition the world's GHG imbalance increases very dramatically. The thawing permafrost also delivers organic-rich soils to lake bottoms, where decomposition in the absence of oxygen releases additional methane that adds to the emission of greenhouse gases into the atmosphere. Local vulnerabilities related to drier landscapes, more wildfires, altered wildlife habitat, and the increased cost of maintaining Alaskan infrastructure are also greatly enhanced by the thawing of the permafrost. (8)

Alaskan communities must rely on improved knowledge of the climate-related changes that are occurring. They must, like communities everywhere, be creative in recognizing, adapting to, mitigating, and responding to the threats and vulnerabilities of a warming climate as they face an uncertain future in which the only sure thing is a continued elevation of their risks and vulnerabilities. Globally speaking, the thawing permafrost represents a major concern for everyone, not just Alaskans.

Hawaii and The Pacific Islands

1. Changes to Marine Ecosystems
   Warmer oceans are leading to increased coral bleaching events and disease outbreaks in coral reefs, as well as changed distribution patterns of tuna fisheries. Ocean acidification will reduce coral growth and health. Warming and acidification, combined with existing stresses, will strongly affect coral reef fish communities.

2. Decreasing Freshwater Availability
   Freshwater supplies are already constrained and will become more limited on many islands. Saltwater intrusion associated

with sea level rise will reduce the quantity and quality of freshwater in coastal aquifers, especially on low islands. In areas where precipitation does not increase, freshwater supplies will be adversely affected as air temperature rises.

3. Increased Stress on Native Plants and Animals
   Increasing temperatures, and in some areas reduced rainfall, will stress native Pacific Island plants and animals, especially in high-elevation ecosystems with increasing exposure to invasive species, increasing the risk of extinctions.

4. Sea Level Rising
   Rising sea levels, coupled with high water levels caused by tropical and extra-tropical storms, will incrementally increase coastal flooding and erosion, damaging coastal ecosystems, infrastructure, and agriculture, and negatively affecting tourism.

5. Threats to Lives, Livelihoods, and Cultures
   Mounting threats to food and water security, infrastructure, and public health and safety are expected to lead to increasing human migration from low- to high-elevation islands and continental sites, making it increasingly difficult for Pacific Islanders to sustain the region's many unique customs, beliefs, and languages. (2014 U.S. Climate Report)

With respect to Hawaii and the U.S. Pacific Islands, the impacts of a warming climate are already being felt in some fairly direct ways. Warmer oceans are leading to increased coral bleaching and disease outbreaks and changing distribution of tuna fisheries. Freshwater supplies are already becoming limited and will become ever more limited on many islands. Coastal flooding and erosion will increase. Mounting threats to food and water security, infrastructure, health, and safety are expected to lead to increasing human migration. (8)

In Hawaii, average precipitation has been declining for nearly a century. For the Western North Pacific, a decline of 15 percent in annual rainfall has been observed in the eastern-most islands in the Micronesia region. A slight upward trend in precipitation has been seen for the western-most islands. On most islands, increased temperatures coupled with decreased rainfall and increased drought will reduce the amount of freshwater available for drinking and crop irrigation. Climate-change impacts on freshwater resources in the region will vary, but low-lying islands will be particularly vulnerable due to their small land mass, geographic isolation, limited potable water sources, and limited agricultural resources. (8)

Rising sea levels will escalate the threat to coastal structures and property. This will have negative impacts on groundwater reservoirs, harbor operations, airports, wastewater systems, shallow coral reefs, sea grass beds, intertidal flats and mangrove forests, and other social, economic, and natural resources. Impacts will vary with location depending on how regional sea level variability combines with increases of global average sea level. On low islands, it is a certainty that critical public facilities and infrastructure, as well as private, commercial, and residential property, are especially vulnerable. Agricultural activity will also be affected, as sea level rise decreases the land area available for farming, and periodic flooding increases the salinity of groundwater. (8)

State agencies in Hawaii have, in response to the accelerating risks and vulnerabilities associated with climate change, begun to develop a framework for climate change adaptation by identifying sectors affected by climate change and outlining a process for coordinated statewide adaptation planning. From an emergency management perspective, this is exactly what should follow on the heels of risk and vulnerability assessment. Indeed, for each region of the United States, it can be said that the risks and vulnerabilities generally summarized here, and backed by solid science as detailed in the Climate Report, should encourage a local response. For each region and each community, the effects of climate change already felt and those to be reasonably projected for the future make abundantly clear the need for all regions and communities to develop a framework for responding and adapting to something that is already happening. Our experiences are telling us something is happening, and the science tells us that what is happening will in all probability get considerably worse.

As we have summarized its findings for each region of the United States, the 2014 multiagency U.S. Climate Report, a report mandated by Congress and developed by hundreds of scientists working with 13 federal agencies, seems to be saying several critical things in the end. It is saying that climate change has moved dramatically into the present. Americans are experiencing it right now with more extreme and variable weather, an increase in intense heat waves and drought, and more frequent and heavy downpours with increased flooding risk. For each region of the country, many of the things people have already begun to experience are the impacts of climate change. It has become a matter of immediate concern to people and the communities in which they live, and each region can expect a future with more extreme and variable weather. As we have reviewed each region and the conclusions of the report, it can be seen that climate change is not just a global issue. It is also, and primarily—from the practical perspective of our personal experiences of it—a very local concern. Connecting sound science with an individual's personal experience, as this report has attempted to do, may be the best strategy to overcome the noise from the corporate disinformation campaign and partisan war that has been waged for the last several decades.

Understanding the Climate "Crisis" in Terms of Infrastructure

For me, the single thing that grabs my attention and sharpens my focus in the U.S. Climate Report is the varying but important impacts of global warming on infrastructure. Each section of the country will see its basic infrastructure stressed and undermined by a warming climate. That makes infrastructure the place to really begin focusing the attention of people where they live. Heat waves, coastal flooding, storm surge, and sea level rise will threaten coastal infrastructure. Decreased water availability in some regions, stresses on electrical grids, and damage to roadways and transportation will be more frequent. Indeed, the 2014 U.S. Climate Report warns repeatedly that U.S. infrastructure is especially vulnerable to the effects of climate change. We must also remember and take note that damage to one infrastructure can often impact other infrastructure systems, causing a cascading effect that spins into multiple widespread effects.

Hurricane Katrina provides an example of this cascading effect. Katrina caused a loss of electricity in the New Orleans region. This meant that several oil pipelines could not ship oil and gas for days and that oil refineries had to be shut down. As a result, gas prices increased dramatically across the country. Hurricane Katrina, whether it was influenced by anthropogenic climate change or not, demonstrates the risks of extreme weather events that exceed the design parameters of our technological infrastructures. A warming climate will only increase the number and the severity of such events. The potential for cascading effects to our regional and national infrastructures may thus be the place where climate change gets really personal for all of us in a hurry. When a major storm wipes out communications lines, a blackout cuts power to sewage treatment or wastewater systems, and people are without electricity, everyone feels it when things begin to mount up. Where an extreme weather event damages a bridge or a major highway, people take note and they are directly impacted. Climate-fueled weather extremes—and many more of these are being projected—will inevitably cause cascading system failures that are fairly common and dramatic unless efforts are made now to minimize such effects. (11)

The Climate Report suggests the threats to infrastructure include more than the physical damage they might suffer in relation to a tropical storm, a flood, or a heat wave. There are additional and often bigger impacts on people, the economy, and the environment that depend on the uninterrupted and smooth functioning of the systems that constitute our basic infrastructure. Disruptions of services in one infrastructure almost always result in disruptions of one or more other infrastructures. This is the cascading effect we have seen and experienced time and time again. During Superstorm Sandy, about 11 billion gallons of sewage was released into waterways when the treatment plants were flooded and lost power. In September of 2011, high temperatures tripped a transmission line near Yuma, Arizona. This sparked a chain of events, a cascading effect that shut down the San Onofre nuclear power plant, caused releases of untreated sewage, and required San Diego residents to boil their drinking water. The power blackout, which lasted some 12 hours, disrupted emergency communications. This made it difficult to notify people that sewage had infiltrated San Diego's drinking water. Altogether some 7 million gallons of sewage had been released, and more than 7 million people lost power. In another incident, a heat wave sparked over 20 different infrastructure failures in a span of 11 minutes that affected millions of residents in Arizona, California, and Mexico. (11) The point is, of course, that climate-related impacts will increase the frequency and the severity of events that impact infrastructures and produce cascading effects.

As the impacts of climate change continue to unfold, close attention will need to be paid in each region and in each community to the risks and vulnerabilities it will pose to our engineered systems for water, energy, transportation, manufacturing, agriculture, coastlines, and other fields. We assume the functioning and the easy availability of our infrastructure. But we often fail to recognize that the infrastructure of the United States was designed to operate within a specific range of climatic parameters and that these parameters are increasingly inadequate as the climate changes. As changing climatic conditions will require our infrastructure to operate outside of the normal range they were designed to operate in, the potential for a series of cascading crises is greatly enhanced. We also might want to remember that even assuming a normal range of climatic conditions, our national infrastructure is a matter of critical concern.

Even without the effects of climate change, the vulnerabilities to America's infrastructures are very large. These vulnerabilities are especially pronounced where they are subjected to multiple stresses and are located in areas that are already subject to extreme weather events. It is also true that breeched levees, fallen bridges, highway disrepair, and weaknesses in an electrical grid badly in need of modernization pose significant problems, even without climate change. This is to say that our aging and crumbling infrastructure is already in crisis without the additional burdens imposed on it by climate change. A good deal of research documents or defines the infrastructure "problem" as a genuine "crisis." (12) The American Society of Civil Engineers, in its report on the state of the U.S. infrastructure, describes what it calls the "serious degradation" of the economy's technological foundations. The deterioration of the U.S. transportation system, for example, has been likened to an iceberg with just the tip of an enormous obstacle to economic growth showing above the surface. The nation's long-term transportation needs, decaying roads, bridges, railroads, and transit systems are considered to be critical. Reports predict dire consequences if the nation does not swiftly address the need to rebuild 60-year-old highway systems and rail lines often far older than that. (12) All of this is without factoring climate change into the mix.

The U.S. Climate Report focuses on near-term effects of climate change on our infrastructure. It tells cities, states, regions, and the federal government what to expect in the next few decades. When one considers the vulnerability of infrastructure because of its aging and disrepair, because of the stress caused by overuse, and adds the impact of their being exposed to new threats like more-frequent and more-extreme weather events, it is easy to see that we have arrived at a critical moment. Climate change makes more urgent the already urgent need to repair, modernize, and upgrade our infrastructures. The urgency to prepare for the effects of climate change on our national, regional, and local infrastructures comes at a time when we have already ignored for too long the need to address an already serious infrastructure crisis. The American Society of Civil Engineers has crunched the numbers and concluded that the United States would need to invest $3.6 trillion to return its aging and broken infrastructure to a state of "good repair" by 2020. (13) There seems to be little chance of coming up with enough public sector financing to meet that need.

In his budget FY 2015, President Obama presented a limited infrastructure proposal to the Congress. Throughout his two terms in office, he has emphasized the need to address infrastructure. In addition to its being an absolute necessity on its own merits, he argued for the benefits that federal investment in infrastructure would mean for the economy and for employment. Of course, Congress has not passed many of his proposals to address infrastructure, as it has been tied up in partisan knots and unable to agree on either the seriousness of the need or how best to approach the infrastructure challenge. Thus, they have been largely unable to legislate anything of major consequence. What the president put forth in his FY 2015 package was very modest. He basically proposed to fund a $302 billion transportation bill for four years. This would fix or repair some roads, bridges, and rails, but it would leave a whole lot more unaddressed. Not included, and not about to be included any time soon, it seems, is funding to attain upgrades or address infrastructure needs in energy systems, drinking water and sewage plants, health care operations, and industrial structures. (11) As modest as this proposal was, the likelihood of its being adopted was probably zero, or very close to it, from the get-go.

Just to add some frosting to the infrastructure cake, one might briefly take note of changes unrelated to climate change that will help to exacerbate its impacts. The population of the United States is expected to rise. It is projected to increase from 310 million in 2010 to over 400 million by 2050. This of course will increase the number of homes and businesses that need things like power, water, and roads. (11, 12) Whether one thinks in terms of our current infrastructure being stressed, aged, and in a state of chronic disrepair and increasing vulnerability or in terms of additional vulnerabilities imposed by climate change or of the impact of population growth, our infrastructure needs are formidable. These needs are quickly escalating, and our vulnerabilities are growing exponentially. Addressing the crisis and looming national disaster associated with infrastructure seems like a sensible proposition, one that should find widely shared agreement from within each region of the country.

In almost every region of the United States, attention to the linkage of infrastructure threats and vulnerabilities to climate change is beginning to be in evidence. Be that as it may, it is still true that adapting infrastructure to the challenges of a changing climate has not received anywhere near the attention it merits. It is almost as if some policy makers and citizens remain largely unaware of the seriousness of the dangers that lie ahead. But the conversation has at least begun in many places. As the realities of climate change and the science that explains it begin to demonstrate conclusively the need to redesign and reengineer our infrastructure to meet new threats and reduce vulnerabilities, the conversation is starting to pick up some steam in communities across the country. A quick snapshot of this conversation may be of interest.

According to the U.S. Climate Report that has been referenced throughout this chapter, 11 of the 12 Northeastern states have developed or are developing statewide climate change adaptation plans. These plans include a variety of mechanisms and strategies to respond to climate change. These include land-use planning, provisions to protect infrastructure, regulations related to the design and construction of buildings, and implications for emergency preparation, response, and recovery. (8)

In the American Southeast, some utilities in the region are already taking sea level rise into account as a critical variable in the construction of new facilities and are seeking to diversify their water sources. The threat of increased inland flooding in low-lying coastal areas and drainage problems already being experienced in many locations during seasonal high tides, heavy rains, and storm surge events are encouraging adaptation options. These include the redesign and improvement of storm-drainage canals, flood-control structures, and storm-water pumps. Adaptations to the threats represented by rising sea levels include community and individual strategies to protect (e.g., building or enhancing levees) and/or adapt (e.g., elevating structures in flood prone areas, wetlands restoration, etc.) to a changing climate. (8)

Climate change presents the Midwest's energy sector with numerous challenges. Its reliance on coal-based electricity and its aging infrastructure mean that a significant investment will be required, even without the effects of climate change, to address the threat of a less-than-reliable electric distribution grid. Renewable energy options may not be as plentiful in this region as in others, but more than one-quarter of national installed wind energy capacity, one-third of biodiesel capacity, and more than two-thirds of ethanol production are located in the Midwest. Increases in precipitation and severe weather will have to be dealt with as well. Some communities have already taken significant steps to reduce future flood damage with land buyouts, and numerous buildings have been adapted with flood-protection measures. (8)

In the Great Plains, the threats presented by climate change cut across sectors: water, land use, agriculture, energy, conservation, and livelihoods. The region's ecosystems, economies, and communities will be further strained by increasing intensity and frequency of extreme weather events. Floods, droughts, and heat waves will penetrate into the lives and livelihoods of Great Plains residents in ever more dramatic and costly ways. Some communities and states have begun to make efforts to plan for these projected changes, but, as is generally and sadly the case nationwide to one degree or another, the magnitude of the adaptation and planning efforts do not match the magnitude of the expected threats. There is, however, according to the National Climate Assessment, tremendous adaptive potential among the diverse communities of the Great Plains. Unfortunately, many local government officials do not yet recognize climate change as a problem that requires proactive planning. (8)

In the Southwest, water is the key problem to be addressed. In California, for example, drinking-water infrastructure needs are estimated at $4.6 billion annually over the next 10 years. This is without beginning to consider the effects of climate change. Climate change is expected to significantly increase the cost of maintaining and improving drinking-water infrastructure, because expanded wastewater treatment and desalinating water for drinking are among the key strategies for supplementing water supplies. The Southwest has perhaps the greatest potential for renewable energy. Abundant geothermal, wind, and solar power-generation resources could help transform the region's electric generating system into one that uses substantially more renewable energy. This transformation has already begun. It is driven in part by renewable energy portfolio standards adopted by five of six Southwest states. From the perspective of the energy infrastructure, this is a very good thing indeed. The projected warming of water in rivers and lakes will reduce the capacity of thermal power plants. Wind and solar installations could substantially reduce water withdrawals. The current levels of demand for water simply cannot be sustained. Thus large water utilities have taken note and are currently attempting to understand how water supply and demand may change in conjunction with climate changes and which adaptation options are most viable. (8)

Among the concerns discussed in the American Northwest, the consequences of climate change will include coastal erosion, inundation, and flooding that will threaten public and private property along the coast. Infrastructure, including wastewater treatment plants, storm water outfalls, ferry terminals, and coastal road and rail systems will be impacted. As a result, municipalities from Seattle and Olympia, Washington, to Neskowin, Oregon, have begun to map risks from the combined effects of sea level rise and other climate factors. Increased risks of wildfires, the possibility of water shortages, and the impacts of drought are also beginning to be factored into state and local plans to adapt to the impacts of climate change. (8)

The vulnerability of U.S. infrastructure to climate-related stress is something we have known about for some time. Research over the past decade has clarified the level of assets at risk from climate change impacts and the necessity for efficient and effective adaptation measures at local, regional, and national levels. (13, 14, 15, 16) These studies have identified the major threats that climate change poses for infrastructures. We have no reason to doubt that the national infrastructure will experience exacerbated damage or destruction from extreme events. Coastal flooding and inundation from raising sea levels, changes in patterns of water availability, and increased maintenance costs associated with higher temperatures in both temperate and permafrost regions are realities that we must understand. (16) Perhaps we do understand these realities in their particularity and as we experience them, but we simply do not easily think of them in relation to climate change or the bigger reality, if you will.

The fact is, we all do understand and have experienced storm damages to or age-related deterioration of roads, buildings, railways, and airport runways. We all understand, and an increasing number of us have experienced, heavy downpours, more severe storms and prolonged rains, intense flooding, and rapid melting of snowpack that exceeds flood protection infrastructure. We understand it is not a good thing if water resources, energy supply, and transportation are increasingly compromised. We understand drought, wildfires, and reduced crop yields. We know about power outages and perhaps even about the need to modernize the electrical grid. But we need to connect these things we understand, know, or have experienced to climate change. It is as if we have not fully grasped the fact that climate change poses a series of interrelated challenges and enhances the threats all of these things we know and experience represent now and for the future. But as extreme weather has become more common over the past several years, engineering specialists have told us that this has amounted to a relentless attack on our infrastructure. They will also tell you that our infrastructure was not designed to deal with such stark changes in weather patterns.

Conclusion

As we have seen, the United States would need to invest $3.6 trillion to return its aging and broken infrastructure to a state of good repair.

Obviously, and without delineating here the politics of it, an intense national effort will be required to address this urgent need. The federal government cannot, even if policy makers were in agreement and inclined to do so, address this need alone. What will be required is a national effort involving national, state, and local policy makers along with engineers, business leaders, and citizens to prioritize and pay for the redesign of some of the most complex infrastructure systems on the planet. The investments needed to restore our infrastructure are large, and the need to make them comes at a time of rapidly accelerating risk and dramatically declining political will or ability to make them. In other words, it will be difficult to strengthen and protect our infrastructure at a time when the political appetite for governing is receding and public resources to promote the public's best interests are in short supply. It is beyond the scope of our discussion in this chapter to resolve this dilemma. But we can, based on the discussion in this chapter, say that the risks and the vulnerabilities imposed by a warming climate require our attention and are of concern to every region, every community, and every individual in the nation.

The vulnerabilities of our infrastructure are due to multiple influences and threats. Among these must be included and managed the impacts of a warming climate. Indeed, a compelling case can be made that the infrastructure crisis in the United States, as documented and articulated by reports and studies generated by the American Society of Civil Engineers, is made more severe and potentially much more costly due to the effects of climate change. As we have examined and summarized the regional assessment of climate change impacts, this conclusion has been reinforced. We are already experiencing climate change in every region of the United States. We are seeing its impact on our local systems, and we are beginning, however reluctantly in some cases, to see that the future will bring more of the same and probably increase the risks to our infrastructures.

Generally speaking, from roads to bridges to utility lines to sewers to coastal protection to agricultural adaptation to flood plain management to freshwater to clean air and to all other infrastructure sectors, we have exacerbated the threats and vulnerabilities we face and must now address. Decades of neglect combined with changing climate circumstances have taken us to the edge of a cliff, so to speak. The cost of this neglect is made even greater by introduction of the new risks and vulnerabilities posed by climate change. Generally speaking, Americans have heard much about their country's infrastructure crisis. A sampling of statements and reports spanning the past decade suggests that this "problem" is old news. The American news media has actually been very consistent in reporting the challenges posed by our crumbling infrastructure. We have been hearing some variation of this message for years, as we see in the following list. The only thing that changes from year to year is the rapidly increasing costs of our inaction.

Selected Media Stories about U.S Infrastructure Challenges

1. "Spending money is part of the solution. China spends 7 percent of its Gross Domestic Product on its infrastructure. India spends 5 percent. The United States spends less than 2 percent. Engineers think the United States will have to spend $2.2 trillion over 5 years to bring the overall grade for infrastructure up to an 'A.'"—Richard Schlesinger, *America's Crumbling Roads and Bridges*, CBS, September 3, 2010

2. "Metropolitan areas tend to have the biggest problems, with 63 percent of urban roads rated in fair, mediocre, or poor condition. Northern cities that are frequently pummeled with extreme weather are especially prone to cracked and eroded asphalt. The longer a pothole or crevice goes unrepaired, the more expensive it is to fix. The cost to resurface one mile of road with asphalt can run about $50,000, whereas major rehabilitation can cost $500,000. Both are better options than starting over completely: one mile of brand new concrete runs from $700,000 to $1.2 million."—Blaire Briody, "America's Third World Roads: Broken and Dangerous." *The Fiscal Times*, June 22, 2011

3. "Here's a sobering statistic: more than 4,400 of the nation's 85,000 aging dams are considered susceptible to failure, according to the Association of State Dam Safety Officials. The problem: the bill for repairing them totals tens of billions of dollars. But not doing so will lead, almost undoubtedly, to economic, environmental, and social devastation. Once again, infrastructure issues rear an ugly head. Who will take care of the technology built by generations before us? And will officials notice before a catastrophe spurs them to act?"—Andrew Nusca, *Five Percent of America's Dams Could Fail*, CBS, February 22, 2011

4. "Several factors are behind the rise. Old dams continue to deteriorate or may fail suddenly because of inadequate spillways and trees growing on dams. Many states don't have enough dam engineers to keep up proper maintenance, causing the

repair backlog to grow. And as more homes and businesses are built closer to dams, the hazards increase, a phenomenon dam-safety experts call 'hazard creep.'"—Mark Clayton, "Problem dams on the rise in US." *The Christian Science Monitor*, September 13, 2007

5. "What will happen if we just continue business as usual? It seems to me that as more and more of transmission infrastructure exceeds its normal life expectancy, there will be more and more blackouts. Areas where there is high congestion, such as the Eastern Interconnection and Southern California, would seem to be particularly at risk. It seems like some of these blackouts could be very long (two weeks?). With the current system, it takes longer to get new transmission lines in place than to build new natural gas or wind generating capacity. Because of this, we are gradually increasing the amount of constriction in the grid. We may have to forgo adding new generating capacity, particularly of wind, until sufficient additional transmission lines can be added. Nuclear plants are big enough that they often can supply power to a fairly large area. If new nuclear plants are added, it may be difficult to add enough transmission lines to use the power they generate optimally. We may find ourselves able to use only part of the power the new plants are capable of generating because of transmission difficulties."—Gail E. Tverberg, "The U. S. Electric Grid: Will It Be Our Undoing?" Energy Bulletin, May 7, 2008

Yes, we have indeed been hearing for some time—decades, in fact—about our pressing infrastructure needs and our growing vulnerabilities. But have we heard the message? We will continue to hear much more of it in 2015, 2016, 2017, and well beyond. The fact is, at least one recent survey of public opinion found that two out of three American voters said improving the nation's infrastructure was very or extremely important to them (74 percent of Democrats, 62 percent of Independents, 56 percent of Republicans, and 59 percent of Tea Party supporters). (17) Infrastructure seems to be a concern that is widely understood. The specific aspects of it that we interact with on a daily basis, the risks and vulnerabilities that we know about and that we have experienced in many cases, and the need to enhance or strengthen it are fairly universally understood. Increasingly, it is even fairly well understood that every dollar we do not spend today, every effort not made today to address the infrastructure problem, will result in the spending of many more dollars and the expenditure of considerably more effort tomorrow. All of this is encouraging, especially in relation to the climate-related risks that will compromise our infrastructure all the more and accelerate the pace of its continued decline if unaddressed.

Adapting our infrastructure to the challenges of climate change is just beginning to receive some attention. The emphasis of both the 2014 IPCC assessment and the 2014 U.S. Climate Report on infrastructure and the risks of extreme weather or climate events to it may well be one of the keys to making the global local and the planetary personal. This conversation, with the emphasis on climate-change impacts as they relate directly to people and infrastructures, may mark a turning point in the climate "debate." Localizing and personalizing the risks and vulnerabilities associated with climate change may well accomplish two absolutely essential things. It will make climate change immediately comprehensible to more people than the science itself has been able to do, and it may create a sense of urgency about responding to it on a policy level nationally, regionally, locally, and personally.

As noted at the beginning of this chapter, an emergency management perspective begins with an assessment of risks and vulnerabilities associated with hazards that may contribute to the potential for predictable disasters. We have seen that every part of the United States is already experiencing climate-change impacts first hand and that the future holds the potential for increasingly severe consequences for communities across the nation. We have seen the need to respond to these risks and vulnerabilities, especially with respect to our various infrastructure systems across the nation, by incorporating the risks and vulnerabilities associated with climate change into our planning activities and the implementation of adaptation measures in each region and each community across the country. As communities across the country begin, as many have in fact begun, to incorporate the risks and vulnerabilities associated with climate change into their assessment of local hazard threats and infrastructural vulnerabilities, it will soon become apparent that every planning activity related to the next flood, the next storm, the next drought, the next wildfire season, the next heat wave, the next extreme weather event of every kind, will be a discussion of climate change. The linkage is unavoidable in a very practical sense. Connecting our practical and local experiences with these various hazards and the risks they pose with the inevitable impact of climate on them is where the global will quite easily become local.

As we now discuss how we might respond to the threats and vulnerabilities we have outlined in this chapter, our focus will shift to the practical steps that our risk and vulnerability assessment must lead to as a logical extension. This leads us to a discussion of the steps necessary to mitigate, adapt, and respond to the threats that have been identified. This means taking proactive steps to reduce the likelihood of the most severe consequences, to reduce the negative impacts, to adapt to the challenges, and to respond effectively to the threats posed by climate change. Just as we prepare for, respond to, mitigate, and adapt to all hazard threats and natural or human-made disasters, we must, after assessing the risks and vulnerabilities, be prepared to respond to global climate change. This response will require global, national, regional, local, and individual efforts.

Once hazard risks and vulnerabilities have been identified, the emergency management perspective turns its attention to the necessary priorities of anticipating and finding solutions to the hazard threats that may contribute to disasters that inevitably lie ahead if we ignore them. This means that the examination of climate change with a focus on changing risks and vulnerabilities to our communities must lead to the management or reduction of the risks as a logical next step. If it does not, then the assessment of these risks and vulnerabilities is a wasted effort. It must be understood, as we move the discussion forward, that there are no responses to risks and vulnerabilities of any kind that are easy or cost-free. As we have seen, the most recent assessments of our infrastructure needs tell us it would cost approximately $3.6 trillion to achieve a state of good repair by 2020. These costs are likely to increase over time. Almost everything we shall be discussing as we turn to the topics of climate mitigation and climate adaptation will involve individual and societal costs that must be borne. At the same time, the cost of doing nothing can be as great, and probably much greater in most cases, than any of the measures we shall discuss. It is also worth noting that the various solutions to the climate crisis will also offer opportunities for economic growth and new development. There are opportunities that will offset some of the costs to be incurred, a notion not discussed and analyzed nearly enough in our public discourse. Whether thinking in terms of costs or benefits, this much can be said to be true. As we consider the risks and vulnerabilities we have articulated in this chapter, *a response is absolutely required.* Whether we act now or later, these risks and vulnerabilities will impose costs no matter what we decide to do or what we refuse to do. If we are smart about it, our response may also create new opportunities. If we are not smart, we will incur greater costs that we will never be able to bear.

References

1. Christoplos, I. (2008). "Incentives and Constraints to Climate Change Adaptation and Disaster Risk Reduction: A Local Perspective." The Commission on Climate Change and Development: Stockholm, Sweden. http://www.gsdrc.org/go/display&type=Document&id=3990 (accessed August 24, 2015).

2. O'Brien, G., O'Keefe, P., Rose, J., and Wisner, B. (2006). "Climate Change and Disaster Management." *Disasters* 30, 64–80.

3. Schneider, R.O. (2011). "Climate Change: An Emergency Management Perspective." *International Journal of Disaster Prevention and Management* 20(1), 53–62.

4. Tompkins, E.C., and Hurlston, L.A. (2005). "Natural Hazards and Climate Change: What Knowledge Is Transferrable?" Tyndall Working paper No. 69, Tyndall Centre for Climate Change Research, University of East Anglia, Norwich.

5. Haddow, G.D., Bullock, G., and Haddow, K.S., eds. (2009). *Global Warming, Natural Disasters, and Emergency Management.* New York: CRC Press.

6. Labadie, J.R. (2011). "Emergency Managers Confront Climate Change." *Sustainability* (3), 1250–1264.

7. Bissell, R.A., Bumback, A., Levy, M., and Echebi, P. (2009). "Long-term Global Threat Assessment: Challenging New Roles for Emergency Managers." *Journal of Emergency Management* 7(1), 18–38.

8. National Climate Assessment (2014). http://nca2014.globalchange.gov/ (accessed July 3, 2014).

9. Intergovernmental Panel on Climate Change (2014). Fifth Assessment report. http://ipcc.ch/report/ar5/index.shtml (accessed July 3, 2014).

10. Resilient Vermont Project (2014). Vermont's Roadmap to Resilience. http://resilientvt.files.wordpress.com/2013/12/vermonts-roadmap-to-resilience-web.pdf (accessed July 10, 2014).

11. Lehmann, E. (2104). "Infrastructure Threatened by Climate Change poses a National Crisis." *Scientific American* March 6. http://www.scientificamerican.com/article/infrastructure-threatened-by-climate-change-poses-a-national-crisis/ (accessed July 10, 2014).

12. McCaffrey, P., ed. (2011). *The U.S. Infrastructure.* Ipswich, Mass.: H.W. Wilson Co.

13. Wright, R. (2014). "US Infrastructure: Broken System." *Financial Times* April 28. http://www.ft.com/cms/s/0/20c50478-ca16-11e3-ac05-00144feabdc0.html#axzz34AjbSCmB (accessed July 10, 2014).

14. Kirshen, P., Matthias R., and Anderson, W. (2006). "Climate's Long-Term Impacts on Urban Infrastructures and Services: The Case of Metro Boston." In *Regional Climate Change and Variability: Impacts and Responses*. Eds. Matthias Ruth, Kieran Donaghy, and Paul Kirshen, 190–252. Northampton, MA: Edward Elgar.

15. Larsen, P.H., Goldsmith, S., Smith, O., Wilson, M.L., Strzepek, K., Chinowsky, P., and Saylor, B. (2008). "Estimating Future Costs for Alaska Public Infrastructure at Risk From Climate Change." *Global Environmental Change* 18(3), 442–57.

16. Neumann, J.E., and Price, J.C. (2009). "Adapting to Climate Change: The Public Policy Response: Public Infrastructure." RFF Report. Washington, DC: Resources for the Future. http://www.rff.org/files/sharepoint /WorkImages/Download/RFF-Rpt-Adaptation-NeumannPrice.pdf   (accessed August 25, 2015).

17. Hart Research Associates (2014). The Rockefeller Foundation Infrastructure Survey. http://www.rockefellerfoundation.org/uploads/files /80e28432-0790-4d42-91ec-afb6d11febee.pdf (accessed July 11, 2014).

CHAPTER 5

Climate Change Mitigation and Adaptation

Introduction

We generally want to assume that we live in a safe world. Some of us like to assume, or so it often seems, that the application of human reasoning will always lead to practical outcomes that will make us the lords of nature. The implicit assumption being that nature is there for our use and domination. Experience and science, of course, combine to repeatedly demonstrate the foolishness of such human assumptions and reveal the fundamental fact that the world is neither safe nor subject to our control. We cannot, as global climate change is reminding us with stunning clarity, control the world and command nature to conform to our desires or meet our needs. In fact, it is we who must adjust our practices and priorities to adapt and respond to a warming climate and the risks and vulnerabilities it will continue to visit upon us with accelerating quickness. Yet we often find ourselves approaching both nature and the future with a sort of hubris that denies our need to be aware of both natural and human-imposed threats to the resilience of necessary natural and human systems.

The impacts of climate change as outlined in the previous chapter and as they are already being felt in every region of the United States and throughout the world present a serious challenge to humanity and to nature. This challenge requires a response. Global warming is a global challenge to be sure, and the response to it must also be global. But as noted in our discussion in Chapter 4, where we sought to view the global in local terms and to personalize the planetary, there is also a critical need to think locally in response to climate change. This is to say that despite the global nature of the challenge and the need for global cooperation in meeting it, local communities everywhere will also need to adjust their practices to cope with, adapt to, or prevent the adverse impacts of a warming climate. A refusal or an unwillingness to do so is a form of hubris we can ill afford, and, it might be added, a display of willful ignorance that is unfathomable in the context of all that is presently known about the very real risks and vulnerabilities associated with climate change. Each community must come to see itself as a primary stakeholder and an agent of change in relation to what we have called a climate crisis. That, for me at least, is the biggest takeaway from our discussion in the previous chapter.

It may be helpful for all of us to think about the climate crisis and the need to respond to it as the reflection of a new normal. Actually, some scientists have coined the phrase "post-normal" to emphasize that point in time when "business as normal" is no longer a satisfactory or workable option. (1, 2) Indeed, some have said that climate change is the perfect example of a post-normal challenge. (3) Post-normal emphasizes that right now is conclusively *not* a time for business as usual. It is no longer adequate. It simply will not work. This implies a need to deal with pressing issues in novel circumstances and in an environment where facts, even where they are not as firmly established by science as is climate change, no longer support the usual expectations or assumptions, values are being debated or disputed by policy makers, the stakes are exceptionally high, and there is an urgent need to make timely decisions. The assumption must also be made, in the post-normal context, that the costs of inaction are as high, even higher in all likelihood, than the costs of action. This seems an apt description of the context for decision making in what we will call the post-normal context known as climate change. Our previous experiences and the working assumptions associated with them are no longer applicable, the post-normal climate is just beginning to be understood, and we need to evolve our understanding to that level where it recognizes the post-normal is actually a new normal to which we have no choice but to respond and to adjust.

The current climatological context, as we have seen, is changing. Previous assumptions about it and about our relationship to it are less and less adequate as we are presented with new risks and vulnerabilities. Having a practical awareness of this by virtue of the solid science that serves as a reliable foundation, and having examined the variety of negative impacts we can expect with some certainty in each region of the country, it makes perfect sense to follow up our discussion of risks and vulnerabilities with a focus on our adaptation to and mitigation of the threats and vulnerabilities associated with a warming climate. That will be the purpose of this chapter. Changing circumstances require adjustments in our thinking and in our practices if we are to adapt to new threats and/or mitigate their impacts. As we shall see, this is a logical next step common in emergency management scenarios, and this logic is directly applicable to climate change and the challenge that it represents. We will begin with a brief overview of some relevant terminology.

Adaptation and Mitigation to the Post-Normal

As we have seen, climate change poses numerous new challenges related to ecosystems, energy, agriculture, weather, infrastructure, health, and communities generally around the country. Many of our previous assumptions that have provided a sense of security, safety, and coherence with respect to the resiliency of our communities and the predictability of our future are being challenged by changing conditions outside of the parameters of our existing structures of thought and activity. In what we might call a post-normal warming climate, many of our conventional assumptions and practices have outlived their usefulness. Under normal circumstances, it might be said that the world makes sense to us. Everything unfolds and is experienced by us in what we perceive to be a coherent and orderly fashion. In a post-normal context, order seems to be replaced by chaos. People begin to feel that the universe, as they experience it, is no longer rational and orderly. When what has been considered normal and orthodox is not working anymore, when it is contradicted by new experiences and changed realities, it is both unsettling and a source of growing conflict and disagreement. Some will recognize and want to respond to changing circumstances, and some will stubbornly refuse to acknowledge them and deny the need to respond differently. Even when the new post-normal climate-related experiences we have call into question the adequacy of our business-as-usual approach, we will be slow to adjust. The post-normal may require significant adjustments in our thinking and in our normal procedures. It may require, in fact, the creation of a new set of procedures and the establishment of a new normal with respect to how we will live our lives. But the temptation is inevitably to keep trying to fit the same old square pegs with which we are comfortable into new round holes where they simply cannot fit. With respect to climate change, we need to stop trying to fit square pegs into round holes. We need to recognize and respond to changing circumstances.

We need to respond to, adapt to, and to mitigate the impacts of climate change with an emphasis on new priorities. We must make an effort to find the appropriate out-of-the-ordinary solutions to new problems. At every level—international, national, regional, and local—we must find practical, responsible, and appropriate ways of dealing with the very real and predictable risks and vulnerabilities associated with climate change. Having examined the risks and vulnerabilities, the next step is to respond. But first, before discussing that next step, let us clarify what is meant by climate mitigation and climate adaptation.

The terms *mitigation* and *adaptation* are familiar to those who work in the field of emergency management. In that field, risk assessment leads to the four phases of disaster commonly described in the emergency management literature. These phases are preparedness, response, recovery, and mitigation. Preparedness consists of the set of activities necessary to respond to natural and human-made disasters that a community is able to anticipate and predict on the basis of thorough risk analysis and vulnerability assessments. This includes the creating of plans, development of capacities, and the acquisition and mastery of technologies and equipment necessary to respond to natural and human-created disasters that a community may reasonably expect to experience on a recurring basis. The response phase, of course, is where the implementation of preparedness plans takes place. This is in the immediate response to a disaster occurrence, and its prime objective is to save lives, assist victims, prevent further damage, and reduce where possible the negative effects of the disaster that has just occurred. Recovery, of course, is the phase where actions are taken in the aftermath of a disaster occurrence, not just to restore or rebuild a community, but to create an even safer and stronger community following a disaster occurrence. These are fairly straightforward and understandable phases in the disaster management cycle. It is the mitigation phase that may be the least understood but most important phase in the cycle.

Disaster or *hazard mitigation* refers to actions taken either before a disaster occurrence or in the immediate aftermath of a disaster occurrence to reduce the likelihood of serious damage losses in a future disaster scenario. In some cases, mitigation may reduce the likelihood of disaster occurrence. It may actually address and seek to reduce the causes of disasters. In other cases, it may lead to adaptations to strengthen the physical infrastructure so that it will tolerate and survive future disaster impacts. Hence, adaptation, which is actually different from mitigation, is considered a part of the mitigation cycle in emergency management. The two terms are often used interchangeably. But strictly speaking, mitigation and adaptation are different sorts of activity. *Mitigation* is more precisely defined as the "reduction of risks." *Adaptation* is defined as the "adjustment to risks" such that a community can successfully cope with their inevitable consequences. The emphasis on mitigation and adaptation in the field of emergency management, both under the rubric of disaster mitigation, is well established as a priority that has a very direct application to what now must be done in relation to the risks and vulnerabilities associated with climate change.

The literature in the field of emergency management talks about an emergency management function not confined to preparing for, responding to, and recovering from specific disasters. Increasingly, emergency management has come to be seen as an integral part of a more comprehensive community decision-making process. It has been increasingly connected to issues such as environmental stewardship, community planning, and sustainable development. (4) More analysis is now devoted to emergency management as a component in broader community planning and development activities. (5, 6) The linkage of hazard mitigation, as a newly enhanced emphasis or priority in the emergency management cycle, to the broader task of developing sustainable communities has placed emergency management at the very heart of community development and planning. (7)

Mitigation, as understood broadly in emergency management, is the assessment and management of hazard risks and community vulnerabilities in an effort to reduce the impact of disasters or, if possible, the likelihood of their occurrence in the first place. The fostering of sustainability in the face of extreme hazard risks and events, natural or human made, is a prominent theme in the current emergency-management literature. It must also be a theme of all efforts we now must make in response to climate change. In fact, emergency-management experience and practice may be said to provide the best guide for how we should proceed. When we speak of climate mitigation or climate adaptation, we are speaking a language with which practicing emergency managers are intimately familiar.

Climate mitigation, as employed in this discussion, may be said to refer to any action taken to eliminate or reduce the long-term risk and the hazard threats that climate change imposes on human life and property. *Climate adaptation* refers to the ability of a system, natural or human, to adjust to climate change, especially to its variability and extremes, in a manner that moderates potential damage and takes advantage of opportunities to minimize them before the situation deteriorates any further. *Mitigation* addresses the *causes* of climate change. *Adaptation* addresses the *effects* of climate change. In general, the more we mitigate, the fewer negative impacts there are to which we will have to adjust. The more preparatory adaptation we undertake, the less stressful and destructive will be the impacts that we do experience. Both mitigation and adaptation will be crucial to reducing our vulnerability to climate change. That is, if we take seriously the risks and vulnerabilities associated with a warming climate, the greatest challenge to be met.

A quick overview of some recent emergency management literature and an examination of some practical responses from the perspective of community-based emergency managers, provide a good sense of the inevitable relationship between climate change threats and the general work of hazard or disaster mitigation. This "snapshot" will provide a perspective on many of the practical concerns stemming from the threats and vulnerabilities presented by climate change. Communities must respond to these threats in the context of future natural disaster scenarios that are clearly implied and which must be managed in the normal course of events even without a warming climate. From there, our discussion will continue with a more in-depth examination of climate change mitigation measures that will be a practical and necessary component in community-based responses to climate change. Local mitigation and adaptation is a practical necessity to create hazard-resilient and sustainable communities in general and especially in relation to climate-related threats and vulnerabilities.

Thinking in Terms of the Next Natural Disaster

Given what we know about global climate change, communities in every region of the United States must be prepared to deal with more frequent and possibly more destructive disaster events. We have seen that weather and climate extremes will be felt everywhere. Communities will have varying and different levels of climate-induced risks to be sure, but all will have risks that become chronic as time goes by. Hazard mitigation techniques that are already employed successfully can teach us much, and they are of course an important building block. Folding climate change assessments into the work already ongoing in communities across the nation to create hazard resilience and sustainable development is doable, and it is a practical necessity. Likewise, policy makers at all levels should promote the necessary connection of traditional hazard mitigation to efforts to measure and anticipate the future impacts of climate change. Indeed, the basic knowledge we already have about climate change should be a strong call to enhance or improve mitigation planning generally. The fact that climate change may increase the frequency and the intensity of expected and regularly recurring events, changing their character and their future impact potential, should provide renewed impetus for mitigation efforts in every community around the globe. Given this, let us take a quick look at the global warming terrain from the perspective of the practical world of the typical emergency management practitioner at the community level.

It is very interesting to note that the general discussion of the relationship between community-based efforts to develop mitigation and adaptation strategies for climate change and its linkage to emergency management is already being commonly discussed. Emergency managers are coming to be regarded as necessary stakeholders and participants, even if not leaders, in such efforts. It is becoming apparent as communities have begun to dialogue and plan in regard to it that local adaptation strategies to climate change are clearly linked to emergency management concerns, especially with respect to disaster risk reduction and hazard resilience. Recognizing this linkage, the Province of Ontario, Canada, for example, included emergency management in its planning process for responding to climate change, saying that it (climate change) was also a critical variable to be factored into the ongoing development of its community emergency management program. (8) An awareness of this linkage has also encouraged many others to specifically mention the inclusion of emergency management in community-based planning groups working on climate change adaptation strategies. (9) There is no disputing the value of including the emergency manager and the emergency management perspective in such planning activity at all community, regional, and even national levels. But beyond such participation, what might the practicing emergency manager incorporate into the day-to-day work she or he performs that would relate to this stated linkage between climate change adaptation and emergency management? The answer to this question may be understood in the context of the four phases of emergency management and the relevance and importance of a warming climate in reshaping the work already being done in each phase.

With respect to the mitigation phase of emergency management, mitigation strategies will have to adapt to the anticipated impacts, short and long-term, of climate change. Coastal communities, for example, may be faced with more frequent and more severe tropical events and the effects of rising sea levels over the next 10 to 40 years. This means that there may be a need for changes in strategies and policies related to land use planning and zoning regulations and the like. There may also be implications for coastal wetlands rehabilitation, flood plain management, business interruption insurance, homeowners insurance, and so on. Severe winter storms, heat waves, wild fires, and tropical events also may impact the transportation and the energy infrastructures and require considerable resources for road repair and power restoration. The point is that such impacts need to be factored into state and local mitigation planning efforts that are already routinely conducted by emergency management agencies. (10)

The rigorous risk assessments that are already a part of the emergency management realm might, as adapting to climate change becomes a necessary part of them in relation to a post-normal climate, suggest the utility of proactive engagement by emergency management practitioners with state and local climate research groups to secure data to support more accurate forecasts of climate change effects. Coordination with local, regional, and state climate change adaptation planning groups would also be advised to support hazard identification and risk assessment. Suffice it to say that hazard mitigation must integrate climate change into the mix of factors that it considers. This applies to all actors, not just emergency managers, who perform tasks and make decisions relevant to the sustainability and hazard resilience of each community.

The disaster-preparedness phase must consider the impact of climate change as well. Preparedness activities will need to account for and adjust to the changing risk profiles associated with climate change. Planning assumptions and scenarios must be re-examined to address the increased frequency and severity of the natural hazards that can be expected to occur within a specific community. Changes in risk profiles may require reassessment of capabilities and a determination about which, if any, additional resources are required for preparedness efforts. Changes in risk profiles may also have implications for new or different impacts on vulnerable populations that need to be considered. In relation to all of these normal considerations, the projected impacts of climate change will render assessments based on past history irrelevant as post-normal climate events change the profile. The incorporation of climate risk and vulnerability projections will be an absolute necessity for any community to be considered to be prepared for the next recurring natural disaster. Disaster preparedness, like mitigation, might benefit greatly from the inclusion of the emergency management community in local, regional, and state climate change adaptation planning groups. (10)

The response phase will obviously be more complex due to climate change impacts for a number of reasons. The impacts will, as we have seen in Chapter 4, be multiple, and they will escalate over time. Local resources may be overwhelmed by more frequent and more severe hazards. As we might easily imagine, the increased magnitude of impacts may require more complex collaboration in response efforts across jurisdictions. Mutual aid among agencies within a community may not, in the face of more frequent events and an escalation in their severity, be able to meet the challenge. Operationally, disaster response may more routinely require more help from a wider array of state and federal agencies. Climate-change impacts will require both a better assessment of larger and more frequent response efforts on local budgets and the anticipation of some command and control challenges associated with larger-scale events that are likely to become more common. (10) New strategies will be required for managing more frequent and complex disasters. The cost of disaster response will continue to increase. This coincides with an era in which governments at all levels, national, state, and local, are unable or simply unwilling to pay for basic services and needs as they resort to budget austerity to reduce what some perceive as the excessive costs of government. This complicates matters significantly.

Needless to say, the disaster recovery phase will also bring new challenges related to climate change. The efficiency of the recovery process may be tested, as it will have to deal with more frequent and more severe disasters. Recovery costs will also escalate considerably. The changing risk profile and the increased frequency and severity of various natural hazards will require emergency management officials and policy makers to make different and more difficult decisions about when to rebuild in certain areas or when to relocate. The question of how to rebuild may also require a different answer, as the rebuilt or new construction will have to be stronger and more resilient.

Mitigation of future impacts must of necessity become a routine part of the recovery and rebuilding process. On a practical level, the creation of decision criteria to guide these decisions in relation to projected climate-change impacts is likely to be a practical necessity.

As one small slice of the pie is quickly examined, from the perspective of the local emergency management function across the country, it is readily apparent that the impacts of climate change are a challenge that must be incorporated into its typical or normal operations. It is only logical, of course, that emergency management is connected to climate change. Most natural disasters that emergency management practitioners must prepare for and respond to are climate related. There are obvious exceptions, such as volcanic eruptions and earthquakes, but every storm, every flood, every wildfire, etc. is climate related. It is also logical to suggest that emergency management and climate change mitigation have common concerns. Both are focused on the protection of lives, livelihoods, and assets. Both are aspects of sustainable risk-awareness development. Disaster risk reduction and climate change mitigation are connected at the hips.

Climate vulnerability, based on knowable current and expected impacts of global warming, is changing. We need to adjust to these changes. What we are calling a post-normal climate requires innovative thinking and adapting to changed circumstances. This means recognition of the post-normal as an indisputable indication of a need to define and understand an emerging new normal and the recognition of the need to respond and adjust to it. That is a practical implication of all that we know about the risks and vulnerabilities identified in relation to climate. As we now move to discuss more broadly what climate change mitigation must logically entail, it will become equally clear that every aspect of local planning, development, and life is implicated. All must be involved if we are to effectively manage the crisis known as climate change.

Climate Change Mitigation

Climate change mitigation, as we have previously stated, may be said to refer to any action taken to eliminate or reduce the long-term risks and the hazard threats that climate change imposes on human life and property. Mitigation, thus, and as we have noted, addresses the causes of climate change. Actually, with respect to climate change, mitigation boils down to one basic thing—reducing the amount of carbon and other greenhouse gases in the atmosphere. Given all that we know about climate change and its causes (Chapter 2), *mitigation* refers to actions that may reduce the human contribution to the planetary greenhouse effect. In other words, the only meaningful mitigation action we can take must be aimed at lowering emissions of greenhouse gases like carbon dioxide and methane and particles like black carbon soot that produce a warming effect.

The fact is that the future of climate change and its impacts is directly related to the choices we will make about greenhouse gas emissions. Lower emissions of heat-trapping gases mean less future warming and fewer severe impacts. Higher emissions mean more warming and more severe impacts. It is also a fact that emissions can be reduced though improved energy efficiency and by switching to low-carbon or noncarbon energy sources. But these facts, indisputable though they may be, are not enough to promote a sense of urgency about climate mitigation. Fossil fuels supply humanity with the majority of its energy needs. Our dependency on them, which seems only to grow, makes mitigation difficult. The power and influence of the fossil-fuel industry effectively applied to the protection of its turf adds significantly to the degree of difficulty. Effective climate change mitigation is possible only with the replacement of high carbon emission intensity power sources (i.e., conventional fossil fuels) with low-carbon sources. This is the only effective mitigation there is, when all is said and done.

The fifth assessment report issued by the International Panel on Climate Change (IPCC) in 2014 shows that the greenhouse gas emissions are still increasing. The science shows that global emissions of greenhouse gases have risen to unprecedented levels despite a growing number of policies and efforts supposedly designed to reduce climate-change impacts. Emissions grew more quickly between 2000 and 2010 than in each of the three previous decades. (11) The U.S. Environmental Protection Agency develops an annual report called the Inventory of U.S. Greenhouse Gas Emissions and Sinks. This report tracks total annual U.S. emissions and removals by source, economic sector, and greenhouse gas going back to 1990. Its 2014 report, released in April and providing a record of U.S. greenhouse gas emission from 1990 to 2012, showed that in 2012, the United States emitted 6,526 million metric tons of CO_2. Total U.S. emissions have increased by 4.7 percent from 1990 to 2012. But 2012 emissions had decreased by 3.4 percent since 2011 and were actually 10 percent below 2005 levels. These changes, considered short-term fluctuations for now, were attributed to multiple factors including reduced emissions from electricity generation, improvements in fuel efficiency in vehicles together with reductions in miles traveled, and year-to-year changes in the prevailing weather. Historically, changes in emissions from fossil-fuel combustion have been the dominant factor affecting U.S. emission trends. The overall consumption of fossil fuels in the United States fluctuates, at least in the short term, primarily in response to changes in general economic conditions. Other factors that may influence consumption rates include energy prices, weather, and the availability of non-fossil alternatives. (12)

The modest decrease in U.S. emissions observed in 2012 is hardly a cause for celebration and, as we have noted, is perhaps a temporary fluctuation. It is entirely possible that as the economy improves, consumption and emission rates will move back upward. In general, and especially relevant with respect to the developing world, economic growth and development leads to higher rates of consumption and emissions. But even if there were to be a permanent drop or the level of emissions were to continue to decline by 20 or 30 percent, it does not solve the problem of excessive carbon emissions. The analogy of an overflowing bathtub is often used to explain why this is so. Consider a bathtub that is full and, as the water continues to flow from the faucet, overflowing onto the floor. Reducing the flow of water from the faucet by 20 or 30 percent, for example, does not solve the problem. Water, albeit at a reduced rate, continues to run into the overflowing tub. The tub continues to overflow, and the overflow continues to cause water damage. We need to turn the faucet off. Either that or we need to pull the plug and drain the tub at a rate faster than the water from the faucet is continuing to fill it up. Given the excess of carbon already in the atmosphere and in the oceans, we cannot reasonably expect to drain the tub (atmosphere) fast enough to stop the GHG overflow. The only sure way to address the cause, to mitigate, if you will, is to turn the carbon faucet off. In other words, we need to stop loading carbon and other greenhouse gases into the atmosphere. While it is almost impossible to imagine us actually leaving fossil fuels in the ground given our present energy economy and the incredible profits it produces for energy producers, this is exactly what we ultimately need to do. This is the only true mitigation measure that will work. Slowing the flow does not stop loading an already overloaded atmosphere. This is not to say we should not reduce the rate of emissions, only that this is not a solution so much as it is a postponement or a delaying of the worst effects of loading up on greenhouse gases.

The largest contributor to U.S. greenhouse gas emissions is carbon dioxide from fossil-fuel combustion. CO_2 emissions from fossil fuels grew by 6.9 percent from 1990 to 2012 and were responsible for most of the increase in national emissions during this period. From 2011 to 2012, these emissions decreased by 3.8 percent. Historically, as we have noted, changes in emissions from fossil-fuel combustion have been the dominant factor affecting U.S. emission trends. (12) Energy generation and transportation are the major sources of fossil-fuel emissions. About 82 percent of the energy consumed in the United States in 2012 was produced through the combustion of fossil fuels. The remaining 18 percent came from other energy sources such as hydropower, biomass, nuclear, wind, and solar energy. (12) Any serious effort at climate change mitigation must begin with an emphasis on reducing the amount of energy produced from fossil-fuel sources and increasing the production from alternative carbon neutral sources. Reduced emissions from electricity generation and improvements in fuel efficiency in vehicles, together with reductions in miles traveled, are among the things that will contribute most to successful climate mitigation.

Even the fossil-fuel companies, always eager to defend their turf, are becoming aware of the need for cleaner-burning fuels. They are even beginning to acknowledge the importance of reducing carbon emissions. At least for purposes of public relations and advertisement, they are beginning to show some awareness. But they are not yet ready to stop expanding the search for fossil-fuel sources or reduce our reliance on them. The boom in natural gas production is being sold to consumers as, in part, a benefit in relation to climate change. New natural gas discoveries, in addition to being plentiful and affordable, are said also to be a bridge to a low-carbon future and a transition fuel that will reduce the negative impacts of climate change. It is true that natural gas is a cleaner-burning fuel, but its benefits in relation to global warming are largely imagined.

A study released in the spring of 2011 called into question the notion that natural gas was a cleaner energy alternative or a benefit in relation to climate change. (13) This study concluded that the greenhouse-gas footprint of natural gas was actually greater than that for conventional gas, oil, or coal. How could this be so? It turns out that when you look at the footprint of natural gas over a longer time span and include in that time span the assessment of waste, leaks, production technology, and consumption, a natural gas well will, over the course of its lifetime, contribute more greenhouse gas emissions than previously thought. In fact, the overall carbon footprint of shale gas extraction can be as much as 20 percent greater than coal, according to this analysis. (13) It turns out that about 35 percent of natural gas fracking wells have methane leaks. The amount of methane that is leaked is much higher than any industrial spokesperson will admit. A growing number of recent studies have also suggested that natural gas, or methane, has a far greater negative global warming impact than previously assumed. Unburned methane, for example, is 33 times more potent in warming the climate than CO_2. (14, 15) The notion that natural gas is the cleaner energy alternative and a bridge to a lower-carbon future in response to climate change may be as much wishful thinking as it is anything else.

It is often the case, unfortunately, that well-meaning efforts to mitigate climate change do not always work out as advertised. Like natural gas and the buzz about it being the "cleaner energy," the deeper one gets into the weeds, the less mitigation we will find. For example, heavy investment in biofuels as a replacement for fossil fuels to mitigate climate change seems like a helpful avenue to explore. But consider that this may provide greater incentives to convert forested land to arable land. Due to their scale, scope, and speed, the immediate effects of these conversions could be more damaging than the direct impacts associated with climate change. With respect to modified landscapes, coasts, and seas, longer-term resilience to climate-change impacts could be significantly reduced. (16) Mitigation may not be as easy to figure out as it might seem.

At a minimum, large reductions in global carbon emissions are generally believed to be necessary if we are to escape the more severe impacts of global warming. Policy actions at the U.S. national level that could contribute to reducing emissions include putting a price on emissions, setting regulations and standards for activities that cause emissions, changing or eliminating subsidy programs, and direct federal expenditures. Proponents of a carbon tax, for example, believe that the price of fossil fuels should account for the societal costs they create. In other words, if you're polluting to everyone else's detriment, you should have to pay for it. In addition to discouraging fossil-fuel use and lowering the rates of carbon emission, the federal revenues generated by a carbon tax could be used to support and incentivize the production of alternative clean energy options. Market-based approaches would include cap and trade programs that establish markets for trading emissions permits. Proponents of a cap and trade system see it as an effective way to deliver significant results with a mandatory cap on emissions without inhibiting economic growth. It is typically said that cap and trade provides polluters flexibility in how they will comply, it rewards innovation and efficiency, and it achieves significant additional emission reductions with minor inconvenience for the polluter. We will examine these options, and others, in greater detail in Chapter 6. For now it is enough to note that none of these measures, and as we shall see they are very tepid measures, have been adopted and implemented at the national level in the United States. The federal government has implemented a number of moderate measures that promote energy efficiency, clean energy, and alternative fuels. These will also be discussed in some detail in Chapter 6, but for now it can be noted here that there are a wide array of very modest national, state, and local actions that are in fact underway to reduce the U.S. contribution to total global emissions. The question is, will they add up to much progress in the end? Let us examine in brief some of these "actions."

Actions and strategies promoted by the federal government are aimed at the reduction of CO_2 emissions from energy use and infrastructure. This makes sense. It can be accomplished through the adoption of energy efficient components and systems, including buildings, vehicles, manufacturing processes, appliances, and electric grid systems. Another strategy has been the reduction of CO_2 emissions from energy supply through the promotion of renewables such as wind, solar, and bioenergy. Other options may include nuclear energy and coal and natural gas electric generation with carbon capture and storage. A third thrust has aimed at the reduction of emissions of non-CO_2 greenhouse gases and black carbon. This might be accomplished by lowering methane emissions from energy and waste, transitioning to climate-friendly alternatives to hydro-fluorocarbons, cutting methane and nitrous oxide emissions from agriculture, and improving combustion efficiency. (17) But as we shall detail in Chapter 6, these efforts have been minimal and inconsistent at the national level. This is related to the partisan divisions in our national government and the resulting inability to enact the legislation that would promote and enable better progress in the reduction of emissions.

Some of the most consistent mitigation activity is being implemented at the state and local level. The federal government refuses to move forward on things like cap and trade, but modest regional cap and trade systems are in place in California and in the Northeast's Regional Greenhouse Gas Initiative. States regulate the distribution of electricity and natural gas to consumers. In fact, jurisdiction for greenhouse gases and energy policies is shared between the federal government and the states. The states, in some cases, are more ambitious than the federal government has been. The most ambitious state policy initiative may be California's Global Warming Solutions Act. Establishing the goal of reducing greenhouse gas emissions to 1990 levels by 2020, this initiative caps emissions and uses a market-based system of trading in emissions credits (cap and trade), as well as a number of regulatory actions. The Regional Greenhouse Gas Initiative (RGGI), formed by 10 northeastern and Mid-Atlantic States, is another multi-state initiative that has been implemented and that has created a very modest cap and trade system applied to the power sector with revenue from allowance auctions directed to investments in efficiency and renewable energy. (17)

An impressive number of voluntary actions aimed at the reduction of CO_2 emissions have been initiated by governmental and nongovernmental entities. Over 1,055 municipalities from all 50 states have signed the U.S. Mayors Climate Protection Agreement. These communities are actively identifying and implementing strategies to reduce their greenhouse gas footprint. Under the American College and University Presidents' Climate Commitment, 679 (including my own) institutions have pledged to develop plans to achieve net-neutral climate emissions through a combination of on-campus changes and purchases of emissions reductions elsewhere. Voluntary compliance with energy efficiency standards and improved building codes developed by industry and professional associations is increasingly widespread. (12) Corporations and businesses, large and small, are beginning to take more small steps to reduce their carbon imprint.

One of the most difficult things about climate change mitigation is that it requires more than public policy and public action of a specific or easily identifiable and universally agreed-upon course of governmental action. It actually requires a societal effort that is all-inclusive. All actors, public and private, have a critical role to play. Successful mitigation, that is, addressing the causes of climate change, will require a profound and determined effort the likes of which can be difficult to imagine. One prominent scholar has noted, for example, that climate change has taken us literally to the end of human history. The meaning here is that humanity has experienced nothing like it in the past, thus our experiences and our standard operating procedures do not begin to prepare us for what is to come next. (18) Another recently published book suggests that climate change "changes everything." (19) Here too the suggestion is that nothing we are currently doing with respect to our economic and political efforts will be able to address the climate crisis which, as both of these authors suggest, may well be an existential crisis for humanity. Without delving into the details of these two treatments or endorsing or rejecting any of their conclusions, the notion that we are facing something very new that requires something very different from us is hard to ignore when we discuss both the arguments for and the techniques of climate change mitigation. In its 2014 report, the IPCC's recommended steps for climate mitigation reinforce the notion. (11)

Outlined below are some of the mitigation measures suggested by IPCC:

- Implement cost-effective fuel switching measures from high-carbon fuels to low- or zero-carbon fuels such as renewables.
- Implement energy-efficiency measures and provide global platforms for energy-efficiency improvement programs.
- Improve existing policies and practices to limit emissions, for example, controlling or eliminating subsidies on fossil fuels.
- Implement measures to raise and expand carbon sinks that trap carbon dioxide, such as forest management and proper land management, etc.
- Improve technology and develop techniques to control methane, nitrous oxide, and other greenhouse gas emissions from the source.
- Promote the use of non-fossil energy sources and conducting research to reduce emissions from existing fossil fuels.
- Revise and implement energy-efficiency standards globally to check emissions.
- Promote environmental education and awareness training in schools and colleges for climate change and associated environmental issues.
- Conduct volunteer programs and form regional action groups to implement climate change mitigation measures.

First, the apparent simplicity of the language and straightforwardness of the IPCC's suggested measures aside, it is strikingly clear that there is no single bullet solution. There is no single optimum technology or a single strategy that can be implemented easily and efficiently to move us from a dangerous and carbon-glutinous energy economy to a carbon-free energy future. Secondly, the number of things that need to be done in a more diversified approach to reduce carbon emissions (policies regulating emissions, the promotion of and investment in clean energy alternatives, improved technology and energy efficiency, regional action groups, volunteer programs, etc.) require concerted and committed action by numerous public and private sector actors, including consumers (e.g., don't buy a Hummer). This can and does in fact work against efficient and successful climate change mitigation.

As climate mitigation efforts may be engaged, and even assuming they are reasonably successful at lowering emission rates, there will still be climate-change impacts in the form of severe weather and environmental threats that must be prepared for and managed. As we have noted, it will be necessary to adjust to climate change, especially to its variability and extremes, in a manner that moderates potential damage and takes advantage of opportunities to minimize them before the situation deteriorates any further. As such, climate change adaptation strategies will be every bit as necessary and important as climate change mitigation. In fact, as we shall see, adaptation strategies are becoming more and more a prominent part of our climate change dialogue in the United States.

Climate Change Adaptation

Adaptation to climate change means taking seriously the predictions scientists have made about its impacts. We have seen that this includes changes in precipitation patterns, heat waves, intensified weather extremes, increased flooding, impacts on agriculture, reduced fresh water, changes in disease patterns, and much more. The only effective mitigation measures to be taken are those that will reduce the loading of greenhouse gases into the atmosphere. This is addressing the causes. But even if we were able to aggressively reduce global greenhouse gas emissions, sea level rise would continue, as would the effects of a warming climate generally. (11) This makes it a necessity to act through conscious adaptation efforts, especially at the local level, to create climate-smart communities that are resilient in the face of mounting threats and vulnerabilities. This does not mean we should be pessimistic about mitigation. In the long run, mitigation or addressing the causes is absolutely necessary, and it will work to reduce the long-term risks. It is simply to recognize that there are effects that we are going to experience even if we do take aggressive mitigation action to reduce greenhouse gas emissions, and these effects must be dealt with as a practical matter by adaptation.

Climate change adaptation means anticipating and understanding the adverse effects of climate change. It means taking appropriate action to prevent or minimize the degree of damage. Well-planned adaptation to climate-change impacts can save money and, more importantly, it can save lives. Examples of adaptation include practical things that have an intrinsic value that is easily recognizable to most people. This includes things such as using scarce water resources more efficiently, adapting building codes to future climate conditions and in anticipation of extreme weather events, building flood defenses, structural reinforcement of threatened infrastructures, developing drought-tolerant crops, choosing tree species and forestry practices less vulnerable to storms and fires, and setting aside land corridors to help species migrate. Due to the varying nature and severity of climate-change impacts from one region of the country to the next, most adaptation initiatives will need to be taken at the regional or local levels.

Examples of communities already undertaking climate change adaptation measures abound. While such efforts are just beginning and many (most) communities are lagging behind in the effort, impressive initiatives are beginning to be observable across the country. Keene, a small city in southwest New Hampshire, is an interesting case in point. Flooding in 2005 caused millions of dollars in damages. With more flooding projected with a warming climate, the city expanded its flood protection efforts to include adaptation to climate change. This adaptation included identifying a 200-year floodplain and prevented future development in these areas to reduce flood risk, assessment of the need for new culvert capacity, identification of existing and future potential animal migration routes to protect wildlife, and the establishment of retraining, scholarship, and loan programs for residents whose businesses are among the most likely to be directly impacted by climate change (such as snowplowing and maple sugar farming). (20)

Another interesting example can be seen in the state of Massachusetts, where the Water Resources Authority incorporated projected sea level rise into plans for building a wastewater treatment plant on Deer Island in Boston Harbor. This plant processes raw sewage and storm water from on-shore communities. Releasing the treated water is done through gravity outflow. This means that the relative level of the treatment facility to the water level is an essential consideration. Projections of a rising sea level had to be considered. Despite the initial additional costs of building the facility 1.9 feet higher than initially planned, this climate-smart adjustment eliminated the need to construct a much more costly seawall and change the discharge processing. (21)

The South Florida Regional Planning Council has created something called a "Climate Change Community Toolbox." (22) This toolbox was developed to help local decision makers, public and private, understand how to carry out adaptation planning and how to conduct analyses of specific sectors. It identifies key areas for focused response strategies and actions, and it presents examples of adaptation planning options. The toolbox consists of three components. Let's look at these components from the perspective of Miami, where the risks and vulnerabilities are quite serious indeed. First, there are general fact sheets that summarize the impacts of climate change on Miami-Dade County's economy, community, and environment. The second component is a sea-level-rise map atlas that contains maps meant to aid Miami-Dade County's Climate Change Advisory Task Force in understanding the magnitude and geography of predicted or expected sea level rise. This is meant to demonstrate the impacts to the county as well the implications these impacts will have for infrastructure and land use planning. The maps also show which areas of Miami-Dade County are generally more vulnerable to sea level rise than others and which areas are likely to experience substantial coastal wetland loss by 2100. This is critical information relevant to planning and development. The third component to the toolbox is a compendium of adaptation resources containing the most current and relevant adaptation processes, strategies, options and examples drawn from national and international sources. (22)

According to the United States Environmental Protection Agency, many U.S. cities have developed or have begun to develop local adaption plans. Several states have also developed or are developing state adaption plans, and regional adaptation efforts also are expanding. (23) The state of Hawaii, for example, in 2014 passed legislation to protect the state against the impact of rising sea levels as a part of its statewide adaptation plan. Hawaii has the fourth largest coastline in the United States and is particularly vulnerable to rising ocean levels. The state has established a climate council (active from January of 2015) to coordinate climate action across all departments within the state government. Hawaii's stated objective is to create plans that will adapt to climate change in both the long and the short term. (24)

The state of Vermont, in the summer of 2014, completed the first state-scale comprehensive assessment of climate change in the United States. This targeted assessment speaks very directly to the impacts of climate change on rural towns, cities, and communities in the state. It also addresses impacts on Vermont's tourism, agriculture, natural resources, and energy. (25)

Some states and localities have had climate change adaptation on their agendas longer than others, but increasingly, all are becoming engaged. In 2011, the state of California was one of the first to produce a climate-adaptation planning guide to support regional and local communities in proactively assessing and addressing the unavoidable consequences of climate change. This planning guide was developed cooperatively by the California Natural Resources Agency and the California Emergency Management Agency. It presents all of the basics for climate-change planning and provides a step-by-step process for local and regional vulnerability and risk assessments. It also identifies adaptation strategies. (26)

Some major U.S. cities have taken the initiative to adapt to climate change as well. The city of Chicago has a detailed climate action plan that includes adaptation strategies to manage heat, pursue innovative cooling, protect air quality, manage storm waters, implement green urban design, preserve plants and trees, engage the public, engage businesses, and plan for the future (see the following list). (27) The goal is to promote adaptation measures that will reduce the impacts and the costs a changing climate will impose on the city. This climate action plan was developed in 2011. New York City developed a similar plan in that same year.

Adaptation Strategies: Chicago Climate Action Plan

1. Manage Heat—Update the heat response plan, focusing on vulnerable populations, complete further research into urban heat island effect and pursue ways to cool hot spots.
2. Pursue Innovative Cooling—Launch an effort to seek out innovative ideas for cooling the city and encourage property owners to make green landscape and energy efficiency improvements.
3. Protect Air Quality—Intensify efforts to reduce ozone-precursors through mitigation programs that reduce driving and emissions from power plants.
4. Manage Storm Water—Collaborate with the Metropolitan Water Reclamation District on a Chicago Watershed Plan that

factors in climate changes and uses vacant land to manage storm water.

5. Implement Green Urban Design—Implement key steps in Chicago's Green Urban Design plan to manage heat and flooding. These steps will enable Chicago to capture rain where it falls and reflect away some of the intensity of the sun on hot days.

6. Preserve Our Plants and Trees—Publish a new plant-growing list that focuses on plants that can thrive in altered climates. Also draft a new landscape ordinance to accommodate plants that can tolerate the altered climate. View the list of recommended trees for Chicago's changing climate.

7. Engage the Public—Share climate research findings with groups most affected—social service agencies, garden clubs, etc. Help individual households to take their own steps to reduce flooding and manage heat waves, such as installing rain barrels and back-up power for sump pumps and planting shade trees.

8. Engage Businesses—Work with businesses to analyze their vulnerability to climate change and take action.

9. Plan for the Future—Use the Green Steering Committee of City Commissioners to oversee City implementation efforts and the Green Ribbon Committee of business and community leaders to assess how the plan is being implemented, recommend revisions, and report to the Mayor and all Chicagoans on our progress. (27)

Climate-related impacts will, as we have seen and repeatedly noted, vary across regions of the country. All U.S. regions will experience some temperature increases, but projected winter warming will be greater in the high latitudes and summer warming will be the greatest in the Southwest. Associated changes in precipitation, winds, and all other climate variables will also vary from region to region. But all regions, all states, and all local communities will benefit from and be wise to invest in the development of adaptation plans. Indications are, as the examples we have just discussed demonstrate, that this is beginning to happen across the nation. Despite the climatological variations that are experienced from region to region, adaptation strategies and actions fall into six basic categories. These include agriculture and forestry, coastal protections, ecosystems, energy and infrastructure, water resources, and human health. (23) The assessment of climate impacts and the risks and vulnerabilities particular to each region or each community will be necessary and critical to help to define the strategic focus and the priorities for climate adaptation efforts in each community.

There are as many ways to adapt to climate change as there are kinds of impacts. Several types of actions are basic and understood as constituting the most commonly used options for adapting. The purchasing of insurance to cover anticipated climate-related losses or disaster costs (i.e., sharing the loss) is an adaptation individuals might reasonably employ, assuming they had enough information to understand the climate-related increases in the level of risk and to anticipate the impacts. Modifying the impact of expected events is another logical adaptation. This might mean constructing levees, sea walls, and storm drainage structures to prevent flooding and anticipate both rising sea levels and increased tropical storm activity. Preventing the effects of climate-change impacts is a widely utilized strategy. This includes improving the ability of physical structures and infrastructure to withstand high winds, intense heat, fire, severe storms, or flooding events, etc. Changing the use of spaces (e.g., convert the use of exposed or high-risk sites from harmful development activities to green space that can absorb impacts or protect communities) and changing the location of communities, structures, and activities that are sensitive to climate impacts to less exposed or lower-risk places are also common adaptation options. (28)

Adaptive strategies may be employed in anticipation of climate-change impacts and in response or reaction to impacts already experienced. The construction of a sea wall, for example, because of future projections related to rising sea levels, is anticipatory. Increasing the capacity of a drainage system after a flood may be both reactive (i.e., the system has proven to be inadequate) and anticipatory (i.e., we will have another flood). Most adaptive measures combine anticipatory and reactive thinking. It should be noted that what we are calling climate change adaptation is anticipatory and/or reactive to climate change as opposed to regular climate variability. This is to say we are not just developing plans to manage the risks of flooding, to use one example, but that we are managing the changing risks or impacts of flooding that are a result of or are directly associated with climate change. It is the current and expected or future impacts of climate change that we are adapting to and that must inform the strategic decisions that provide the focus for all efforts. Likewise, adaptation requires the integration of efforts by state and local agencies, public and private businesses, nongovernmental organizations, and individuals to prepare for, address, and adapt to the impacts of climate change.

The basic principle of climate adaptation is to integrate it with a community's or organization's overall strategic planning. In other words, climate risk must be mainstreamed as a normal component in all risk and vulnerability assessment and planning. Who needs to be involved in every community in climate adaptation planning? The short answer is everybody. The long answer is public agencies, business associations, major companies, and nongovernmental bodies with responsibilities and interests in water and energy resources, coastal development, agriculture, food processing, fisheries, forestry, human health care, infrastructure and the constructed environment, biodiversity and habitat, tourism, natural disaster management, and all businesses and citizens who will be impacted. How does a local community or a state go about structuring a planning process for climate adaptation? Based on the examples and experiences of communities already engaged, a pretty good road map has been established. The basic steps in local and regional climate adaptation planning include the development of an organizational structure for planning. This is followed by a planning process that assesses likely climate impacts, risks and vulnerabilities, and response capacities, and develops a strategic plan for adaptation, implementation of the plan, and evaluation of performance and changes in risk patterns. Let us examine this process in a bit more detail.

The first step commonly taken is to identify the agency or organization that will act as the home base for the adaptation process. This may be an existing entity or a newly created one with a targeted focus on climate change and all planning pertaining to it. A set of stakeholder organizations from public and private sectors is generally recruited to partner in the planning effort. As a part of the overall effort, the planning body needs to develop channels of communication and action to involve people and stakeholder organizations in the planning and action processes of adaptation. This usually includes, throughout the planning and implementation processes, things like a website, a speaker's bureau, town meetings, the involvement of schools and colleges, and regular media briefings.

The planning for climate-related risk adaptation begins with an assessment of climate-change impacts that a community can expect and is, in fact, already beginning to experience. The purpose here is to identify and prioritize likely impacts of climate variability and change that will either amplify current stresses or create new ones. This, naturally, requires information. An assessment of climate-change impacts will require interdisciplinary teams of experts and stakeholders who are capable of conducting a scenario planning process based on regional or local applications of climate change models. This assessment can also be enriched by stakeholders who are able to identify climate impacts based upon their working experience and the local or regional history of climate events and trends. In the assessment of risks and vulnerabilities, it is necessary to create alternative scenarios that evaluate best-case, middle-case, and worst-case possibilities. But the subsequent development of adaptation goals and strategies must, in addition to addressing current or already experienced effects of climate change, strive to address the most likely future trends of climate change.

In developing the actual strategic plan for climate change adaptation, it will be critically important and useful to link strategies to other important trends such as the transition to sustainable farming, energy and water efficiency, the implementation of renewable energy technologies, and sustainable land use master plans that can enhance the effectiveness of climate adaptation. All adaptation strategies must be prioritized according to the likelihood, intensity, and potential cost of the climate impacts they are meant to address. The ultimate purpose is to choose policies, resource investment strategies, and action plans that address the local or regional effects of climate change in an effective and affordable manner. It is often suggested that the best strategy, at least to begin with, is to choose and implement strategies and responses that will pay off or constitute a benefit to the community, no matter how the impacts of climate change eventually play out.

Once the adaptation strategies and the plan of action have been formulated, they must be implemented. Responsibilities for implementation must be clearly articulated and specific agents identified. The organizational structure and the communication channels must also be designed to assure clear responsibility for oversight and coordination of stakeholders and participants. Naturally, the implementation of the adaptation plan must be routinely evaluated. In addition to monitoring performance and upgrading it where necessary, changes in risk and vulnerability patterns must be incorporated into any adjustments found to be necessary.

As noted, it is suggested that it might be wise to choose and implement strategies and responses that will pay off or constitute a benefit to the community no matter how the impacts of climate change eventually play out. The initial steps in climate change adaptation may be easily seen as extensions of already ongoing and necessary activities. These are often called "no-regrets" actions. An example of this can be seen in relation to the potential goals or strategies that would be most relevant for coastal communities in relation to climate change. Many of the things that would be considered in that connection are items that would, in one way or another, already be on the agenda of a coastal community. The National Oceanic and Atmospheric Administration (NOAA) provided an example of goals relevant for coastal areas as they adapt to climate change. (29) Based on what is known about the effects, (current and expected) of climate in coastal regions, each of the goals articulated will pay off if they are achieved (see the following list). Enhancing the resiliency of natural and human systems to deal with stress is a good thing, even if the stress is not climate related. Preserving and protecting or rehabilitating natural coastal systems is a good thing. These things are of value and constitute a benefit to coastal areas regardless of a warming climate and a rising sea. But because climate-change effects in coastal regions are not a matter of dispute, because they are happening and will continue to do so, any actions that reduce further damage and that enhance adaptive capacity will be no-regrets actions. The concept of no-regrets action is commonly used, as already noted, in reference to actions that will have a beneficial impact regardless of what the future will hold with respect to climate-related risks and vulnerabilities.

NOAA Climate Change Adaptation Goals for Coastal Communities

1. Reduce the vulnerability of the built environment to sea level rise.
2. Monitor and maintain functioning and healthy coastal ecosystems.
3. Reduce the costs associated with disaster response and recovery.
4. Protect critical infrastructure from the impacts of climate change.
5. Minimize economic losses attributable to the impacts of climate change.
6. Adapt to climate change in a manner that minimizes harm to the natural environment and loss of public access.

7. Increase public awareness about climate change and how it affects the coast.
8. Reduce the impact of climatic and nonclimatic stressors on natural systems. (29)

Successful adaptation to climate change begins with the recognition that adaptation absolutely must happen at the local level. Each community must identify key vulnerabilities (exposure, sensitivity, adaptive capacity). They must involve all key stakeholders, set priorities for action based on projected and observed impacts (magnitude, timing, persistence, likelihood, costs, human effects, economic effects), and they must choose adaptation options based on a careful assessment of efficacy, risks, and costs. This point is stressed in the 2014 U.S. National Climate Assessment. Recognizing that adaptation to climate-change impacts on human health, natural ecosystems, built environments, and existing social, institutional, and legal arrangements are a necessity, the National Climate Assessment emphasizes the practical value of local adaptation. It notes, for example, that building codes and landscaping ordinances will likely need to be updated not only for energy efficiency but also to conserve water supplies, protect against disease vectors, reduce susceptibility to heat stress, and improve protection against extreme events. (30) The report also recognized some challenges that may need to be overcome before local climate change adaptation might be as successful.

The National Climate Assessment notes that climate change adaptation actions do often help to fulfill other societal goals, such as sustainable development, disaster risk reduction, or improvements in quality of life, and can therefore be incorporated into existing decision-making processes. This is a good context for what is typically called no-regrets action. The report notes that, as we have seen in our discussion of several examples, substantial adaptation planning is occurring in both public and private sectors and at all levels of government. But it also notes, with some notable exceptions in a few localities, few measures have been fully implemented and there is a need to pick up the pace if we are truly to adapt to climate change. (30)

Approximately 60 percent of local communities (at the date of this writing) across the United States are engaged in some sort of climate-change adaptation planning. In addition to local government planning, regional agencies and regional aggregations of governments are becoming significant climate change adaptation actors as well. But the planning process has just begun in many communities, and the implementation of these plans has yet to occur in many cases. Most of the adaptation that has actually been implemented to date has taken place at the local level. Again, the basic mechanisms include land-use planning; provisions to protect infrastructure and ecosystems; regulations related to the design and construction of buildings, roads, and bridges; and emergency preparation, response, and recovery. (30) But it must be emphasized that the process is far from complete. It must also be a priority to begin planning in those communities that have not yet participated.

The U.S. national government has initiated some measures to promote and support climate change adaptation as well. Several new federal climate-adaptation initiatives and strategies have been developed in recent years. Most of these are the result of executive orders issued by President Obama, as the U.S. Congress has shown little to no willingness or ability to legislate on climate change. Key federal initiatives include the items mentioned in the following list as they appear in the National Climate Assessment.

Federal Climate Change Adaptation Initiatives

1. Executive Order (EO) 13514, requiring federal agencies to develop recommendations for strengthening policies and programs to adapt to the impacts of climate change;
2. the release of President Obama's Climate Action Plan in June 2013, which has as one of its three major pillars preparing the United States for the impacts of climate change, including building stronger and safer communities and infrastructure, protecting the economy and natural resources, and using sound science to manage climate impacts;
3. the creation of an Interagency Climate Change Adaptation Task Force (ICCATF) (now the Council on Climate Preparedness and Resilience, per Executive Order 1365368) that led to the development of national principles for adaptation and is leading to crosscutting and government-wide adaptation policies;
4. the development of three crosscutting national adaptation strategies focused on integrating federal, and often state, local, and tribal efforts on adaptation in key sectors: 1) the National Action Plan: Priorities for Managing Freshwater Resources in a Changing Climate; 2) the National Fish, Wildlife and Plants Climate Adaptation Strategy; and 3) a priority objective on

resilience and adaptation in the National Ocean Policy Implementation Plan;

5. a new decadal National Global Change Research Plan (2012–2021) that includes elements related to climate adaptation, such as improving basic science, informing decisions, improving assessments, and communicating with and educating the public;

6. the development of several interagency and agency-specific groups focused on adaptation, including a "community of practice" for federal agencies that are developing and implementing adaptation plans, an Adaptation Science Workgroup inside the U.S. Global Change Research Program (USGCRP), and several agency-specific climate change and adaptation task forces; and

7. a November 2013 Executive Order entitled "Preparing the United States for the Impacts of Climate Change" that, among other things, calls for the modernizing of federal programs to support climate-resilient investments, managing lands and waters for climate preparedness and resilience, the creation of a Council on Climate Preparedness and Resilience, and the creation of a State, Local, and Tribal Leaders Task Force on Climate Preparedness and Resilience. (30)

All federal agencies are required to plan for climate change adaptation. This includes coordinated efforts by the White House, regional and cross-sector efforts, agency-specific adaptation plans, and support for local-level adaptation planning and action. (30)

Despite the growing involvement of local and regional entities in planning efforts and the initiatives at the national level, the National Assessment concludes that many barriers impede climate change adaptation efforts. These barriers include:

> . . . difficulties in using climate change projections for decision-making; lack of resources to begin and sustain adaptation efforts; lack of coordination and collaboration within and across political and natural system boundaries as well as within organizations; institutional constraints; lack of leadership; and divergent risk perceptions/cultures and values. (30)

We will discuss a few of these barriers in greater detail in the next chapter. It must also be remembered that our success or our failure with respect to climate change mitigation will either increase or decrease both the cost and the effectiveness of our efforts to adapt to a warming climate.

Despite the positive indications that reductions in carbon emissions are recognized as being essential to mitigate global warming, and despite the growing number of state, regional, and national efforts to adapt to climate change, there are still too many examples of efforts to block or retard such efforts. The efforts to adapt to a warming climate are encouraging, but they do not yet add up to anything near what is needed. Too many examples of a refusal to adapt to a warming climate can be found. One of these is so foolhardy, and yet instructive, it is worth taking note of here as an indication that progress cannot be assumed to be steady across the nation.

The Republican-controlled legislature in North Carolina, in a bizarre example of resistance to both climate adaptation and climate mitigation, passed a law banning the state from basing coastal development and redevelopment policies on the latest scientific projections of how much the sea level will rise. The law was crafted in response to an estimate by the state's Coastal Resources Commission (CRC) that the sea level rise would be 39 inches in the next century. Coastal developers, partnering with climate change deniers, brought pressure to bear. This legislation was prompted by fears of costlier home insurance and accusations of antidevelopment alarmism by residents and developers in the state's coastal Outer Banks region. (31) The state's General Assembly thus said that the scientific projections made by climate scientists regarding sea level rise are to be disallowed or ignored by state agencies. Instead, projected sea level should be based on historical record instead of science. This new state law mandates that projection models use a linear increase (i.e., a consistent amount of change every year, based on historical data) instead of the projections based on climate science. This will lead to predictions that are, of course, much less catastrophic and much more reassuring for people building resorts in the Outer Banks of North Carolina. These predictions will also be flat-out wrong. (31) But being right, or scientifically correct, in North Carolina is now apparently against the law.

The North Carolina story is interesting on several levels. It exemplifies the contest we discussed in Chapter 3 between science and everything not science. Contrary to the North Carolina law, actual sea level rise is nonlinear because there is feedback. As we saw in Chapter 2, the warmer it gets, the more the water volume expands. More ice melts, warming water expands, and the sea level rises. But North Carolina legislators didn't like the science, so they passed a law saying the scientific result was, so to speak, disallowed. But why pass such a law? Climate change denial is only a part of the answer. In fact, it is probably a very small part of the answer. The ultimate answer begins with a look at a projection map showing land along the coast underwater. This scientific projection, if accepted, would place the permits of many planned development projects in jeopardy. Numerous new flood zone areas would have to be drawn, new waste treatment plants would have to be built, and roads would have to be elevated. The endeavor would cost the state hundreds of millions of dollars. In addition, business and property owners worried about property values and resale potential could see in that projection map threats to their economic interests. The economic consequences of preparing the North Carolina coast for a one-meter rise in sea level (under which up to 2,000 square miles is expected to be threatened) were unacceptable to the proponents and supporters of the bill. Better from their perspective to question the science, join with deniers in a marriage of convenience, or simply become a denier out of expedience, to forge a policy more friendly to their immediate economic interests. But the costs of addressing the reality of sea level rise that are regarded by developers and deniers as excessive and damaging to the economy are likely to be very small when compared to the costs of failing to mitigate or adapt to the negative effects of climate change. This disagreement with and disregard of science by the North Carolina General Assembly pretty much summarizes the political contest about climate change over the past several decades. This contest rages on at the national level. Stunningly, and even more absurdly, in the spring of 2015, a couple of states officially banned the mere mention of the words "climate change" by state officials. One of these states, Florida, is ground zero for the impacts of rising sea levels.

With respect to both climate change mitigation and climate change adaptation, progress on the policy front has been slow. The United States Congress has never enacted greenhouse gas legislation. It shows no signs of doing so any time soon. Reviewing the proposals that have been made (but not enacted) by both the U.S. Congress and by international bodies that are aimed at reducing greenhouse gas emissions is not encouraging either, at least according to some. The problem with the proposals made, according to their critics, is that they are all simply a matter of too little too late. They promote a gradual approach to the reduction or abatement of greenhouse gases, and, as such, would be (even if adopted and implemented) ineffective because greenhouse gases would continue to accumulate and build to higher levels, thus contributing to the warming of the climate. In other words, most national and international proposals are inadequate because they back-load the problem toward later decades rather than advance a solution in a timely and efficient manner. (32) Even as nations are beginning to show a greater sense of urgency about reducing emissions, there is a fear that this has happened too late in the game.

The problem with conventional mitigation approaches such as cap-and-trade systems or carbon taxes, beyond their being impossible to enact, it seems, is they really don't mitigate very much at all. A cap-and-trade regime sanctions emissions according to allotted emissions allowances. But the worst corporate emitters can influence allotments such that emissions are not very effectively reduced when all is said and done. A carbon tax increases the price of fuel and theoretically discourages use and incentivizes the development of replacement renewable energy sources. But ultimately, both continue to add greenhouse gases to the atmosphere. The net result, at least according to some, is that neither a cap-and-trade nor a carbon tax will successfully reduce fossil-fuel use or carbon emissions soon enough to make a positive difference. (32) Even though it has proven difficult to adopt and implement these fairly tepid conventional mitigation measures, which may not mitigate much at all in the end, a consensus seems to exist that such measures are the most practical and likely means available to build a foundation for national and global cooperation in the reduction of emissions. The fly in this ointment, again according to its critics, is that this consensus is a back-loaded approach that continues to emit and accumulate higher levels of greenhouse gases into the atmosphere and thus leads to a worsening condition with respect to global warming. The assumption implicit in the market strategies and carbon pricing schemes is that we can reduce emissions while still accelerating our extraction of fossil fuels and that a gradual shift to cleaner energy will be possible without any significant changes in the way we do business in our energy economy. Critics of these approaches suggest that we cannot afford to wait for the ultimate failure of such approaches to become evident before we acknowledge the need for more rapid and fundamental change. Those who adhere to this view regard matters to be serious enough to require a rapid transition to carbon-free technologies. They emphasize that today, not tomorrow, is the time to develop ways of obtaining energy that eliminates the release of CO_2 and other greenhouse gases.

In other words, the only way to reduce the greenhouse gas levels in the atmosphere enough to reduce (or even survive) climate-change hazards is to replace large pollution sources as rapidly as feasible in as many industrial sectors and geographic regions as possible with clean alternative technologies, processes, and methods.

The lack of progress in lowering emissions rates, the slow rate of progress on renewables, and spreading concerns that we may not be able to move fast enough on these fronts to manage the climate crisis has led to more discussion of new technologies to intervene in nature to counter the negative effects of climate change. The National Academy of Sciences, NASA and other governmental agencies, and even the U.S. Intelligence community all agree that drastically reducing emissions of carbon dioxide and other greenhouse gases was by far the best way to mitigate the effects of a warming planet. But they share concerns that we will simply not make progress in doing that. Thus there is growing discussion and consideration of the most dramatic and dangerous of technological adaptations—geo-engineering. (33)

In February of 2015, a National Academy of Sciences panel recommended, somewhat reluctantly, that it might be necessary to do more research on geo-engineering responses to climate change. Geo-engineering options consist of efforts to either capture and store some of the CO_2 that has already been emitted, to reduce the greenhouse effect, or to reflect more sunlight away from the earth so there is less heat to start with. CO_2 removal would be very expensive if pursued on a planet-wide scale, would take many decades to implement (too long, most likely, to avoid the worst impacts of climate change), and would most likely not keep the CO_2 out of the atmosphere permanently. Solar radiation management, or reflecting sunlight away, is very controversial and potentially dangerous. Most discussions of this alternative focus on the idea of dispersing sulfates or other chemicals high into the atmosphere. This would reflect sunlight, in some ways mimicking the effect of a large volcanic eruption. This would no doubt lower temperatures fairly quickly, but the process would have to be repeated frequently. This, in turn, will have unintended effects on weather patterns around the world (bringing drought to once-fertile regions, for example). It might also be used unilaterally as a weapon by governments or even extremely wealthy individuals. Opponents of geo-engineering, and this includes many of the scientists who reluctantly agree that these options need to be explored, argue that even conducting research on this subject presents moral hazards and will succeed primarily in distracting society from the absolutely necessary task of reducing the emissions that are causing warming in the first place. (33) Interestingly, some of the largest investments in geo-engineering research are being made by many of the same corporations and front groups that have been spending millions and millions of dollars to deny that climate change is happening in the first place. That's right. They are investing in a solution to a problem they say does not exist. This means they sense the day will come when denial won't work anymore, and they want to be ready to dictate "solutions" in a manner that keeps fossil-fuel extraction going full speed ahead and the rolling in of its dirty profits uninterrupted. (34)

While the progress on climate adaptation has been uneven and incomplete, it is true that some small progress has been observable and much more is possible with respect to adaptation efforts, especially at the local level. But this progress to date has been very small and does not begin to make enough of a dent. Climate-mitigation efforts have not been nearly as impressive. In fact, based on results to date, it is widely believed that there is little to no chance for serious national or international policy initiatives to be adopted and implemented to cut emissions nearly enough on the near horizon. This leads many to the conclusion that mitigation is a failure and adaptation is our only remaining option. Indeed, it has become almost alarming to observe the growing number of people who say mitigation is a lost cause. This is to suggest that all we can hope to do is adapt to a warming climate. Even more alarming is the renewed consideration, however reluctantly, of the geo-engineering solutions that may pose more new threats to humanity that outweigh their benefits in managing climate-change impacts. But if we refuse to mitigate or are convinced that mitigation will not be politically feasible, adaptation generally and geo-engineering in particular, with its unstoppable path to full-scale deployment of new but risky technologies, will appear as the only viable options. That would indeed be unfortunate at best and a disaster at worst.

To Adapt or to Mitigate: Is That a Question?

What is the best way to address the risks of climate change? Should we mitigate or should we adapt? Many people are having this conversation as though it were an either-or proposition. Should the world cut greenhouse gas emissions to lower the risks of harm from climate change (mitigation), or should we just get used to it (adaptation)? Should we work to decarbonize our energy economies or should we be spending money to build seawalls, move populations inland, and figure out how to grow food in a world of higher temperatures, persistent droughts, and extreme weather? My answer to all of these questions, with my emergency management mind-set, is "yes." We need to adapt *and* to mitigate. This answer is pragmatic. Adaptation is absolutely necessary, and resistance to mitigation is futile.

Updating flood maps, building sea walls, reinforcing threatened infrastructures, developing drought-tolerant crops, and other adaptive measures are very necessary and practical steps to take. As we have seen in our brief overview of adaptation, communities will have expected and increasingly negative impacts. But these impacts can be managed, and damages, as well as human suffering, can and should be reduced with intelligent forethought, planning, and action. The fact that mitigation is more difficult, that the politics of it is polarized to the point of absurdity, that progress is crippled or retarded by our short-sighted partisanship, and that it will just take longer even under the best of circumstances means that there will be an ever-increasing number of negative impacts we must be prepared to deal with for some time to come. Adaptation serves that purpose and is a practical necessity. But one must hasten to add that adaptation is not enough to deal with climate change. It is not the solution to the crisis, nor can it prevent an ultimate disaster.

We might consider climate adaptation as a sort of Maginot line in relation to climate change. Adaptation is fortification; it blunts the attack, it buys us some time, and is intended to protect us against the very worst as we prepare to mobilize and respond to climate change. It does not, however, make us impervious or immune to all of the negative impacts or suggest that we can continue to be lax with regard to climate mitigation. Neither sea walls nor geo-engineering will prevent ocean acidification, water shortages, more severe droughts, more costly and dangerous wildfires, declines in agricultural output, or many other impacts of climate change such as the inevitability of climate refugees and greater risks of war in regions most severely impacted and challenged. Mitigation is a practical necessity to deal with the causes of climate change. Adaptation has its limitations, and it cannot exceed them. At some point, we must recognize that there's no adapting to the changes in the climate we are facing if we don't cut emissions dramatically. At some point, the continued dumping of greenhouse gases into the atmosphere will create a situation in which adaptation is no longer possible. The time that adaptation will buy us is not unlimited, and the cumulative effects of our failure to mitigate, to turn the carbon faucet off, will eventually overrun our adaptations. My emergency management orientation reinforces my conclusion that we must do both—mitigate *and* adapt.

From the perspective of emergency management, once we have assessed the risks and vulnerabilities, we are in a position to identify various hazards and to anticipate future disaster impacts. Emergency managers generally look at the world differently. They understand that natural disasters, for example, are not the work of nature alone. They are instead the predictable results of interactions of the natural environment, or the earth's physical systems, with the social, economic, and demographic characteristics of the human communities in that environment, and the specific features of the constructed environment (i.e., buildings, roads, bridges, and other features). In this same vein, the impacts of climate change are predictable. As our discussion in Chapter 4 demonstrated, all communities across the United States, and around the world for that matter, need to consider climate change projections relevant to their specific location and to factor these into their decision making. The predictable results of interactions of a warming climate with the social, economic, and demographic characteristics of the human communities and the specific features of the constructed environment will mean that all traditional functions (including emergency management) will of necessity have to incorporate climate change adaptation into their normal workload.

Emergency management professionals have already incorporated the adaptation to hazard potentials and disaster threats into their work. A normal part of their task is to anticipate and to take steps to reduce severe impacts and to protect the resiliency of the communities they serve. Likewise, they have incorporated hazard mitigation into their work. The linkage of hazard mitigation to the broader task of developing both hazard-resilient and sustainable communities has placed emergency management and the work associated with hazard mitigation and adaptation at the heart of development and planning in every community. With respect to climate change generally, both mitigation and adaptation must be incorporated into the mix if communities are to respond to its threats. Emergency management practice offers a perfect model for the logic of this statement.

Adaptation in the form of better engineering, improved construction, and fortified infrastructure is a proven wise investment in that it can significantly reduce the impacts and costs of anticipated or predictable disasters. But as emergency managers know, adaptation has its limits. Adaptation may reduce impacts, but it cannot eliminate them. Also, adaptation sometimes fails. Flood control structures, for one example, not only frequently fail to meet the standards of reliability set by planners, but can, in some cases, even increase damages in the long run. This is to say that structural adaptations may increase damage where they lull a community into a false sense of security that results in their continued and unwise investment in the development of high-risk locations. The false assumption that structural design or adaptation alone is sufficient to prevent future damages may lead to an under-emphasis on mitigation. The result might be that the adaptation can successfully delay or postpone damages, but when the eventual failure or limitations of the adaptive strategies and structures are reached, it may actually set a community up for greater and more costly damage at a later date. It is for just this reason that emergency management practice does more than manage for or control for the effects of disasters (i.e., adaptation). It emphasizes strategies to address the causes (i.e., mitigation) as well.

Emergency management spends a great deal of its time and effort on adaptation, but recognizing its limits, it also emphasizes hazard mitigation. In addressing causes of natural disasters, there is no suggestion that humans can control nature or prevent its inevitable extremes. Rather, and very importantly, it is to suggest that humans can play a role in preserving and protecting natural ecosystems and in promoting the sustainable use of the earth's resources. Such stewardship is a necessity and a primary means of mitigating the threats posed by natural hazards. It is necessary to ensure that our communities will be both resilient and sustainable. This is to suggest that human beings, and the choices they have the ability to make about where to live, how to live, and how human and economic development will proceed, determine the severity of the threat and the losses that their communities will bear in future disaster scenarios. As with natural disasters in general, it can be said with respect to global warming that the choices we make today will shape the nature, impact, and severity of climate change tomorrow. Mitigation is all about making better choices. It begins with the insight that the resilience and the sustainability of our communities requires that we understand, respect, and protect the natural order by intelligently accommodating ourselves to its predictable and inevitable extremes. A failure to do this or a conscious choice to refrain from doing it only results in catastrophe and losses that are entirely the product of human design.

It is no doubt generally true, and especially so with respect to climate change, that adaptation is easier to promote than mitigation. It focuses on more immediate concerns connected to nearer-term events. It is especially easy to see the value of strengthening infrastructures or flood control measures in relation to property and the physical or constructed environment. But talk of climate mitigation may be more difficult. We do not as easily think about or perceive ecosystems. Even if we understand that ecosystems are threatened, we are less inclined to see an investment in reducing that threat as wise in the shadow of all of the more immediate economic and practical concerns more easily recognized before our eyes. We may be inclined to address what we see as more relevant based on past experience and reasonable expectations (storm damages, rising seas, flood risks) but dismissive of things not so directly a part of our experiences and expectations (CO_2 levels). If we can't see it or touch it, we might feel it is less important in the immediate scheme of things.

The time frame also complicates matters. Adaptation can be tied to events that have already occurred and/or things on the immediate horizon. There will always be a next storm. A documented rise in sea level and projections for future rising of the sea level can be understood in a more immediate context, even if the worst of the problem is far off into the future. Be that as it may, we might have already experienced an increased or more dramatic storm surge and thus see the need to address it. Sea walls make sense. But if we are talking about impacts of rising CO_2 levels and discussing how to mitigate greenhouse gas emissions, it is a very different proposition. The most severe effects are seemingly far off. It is incredibly easy to ignore the effects we already feel, because CO_2 emissions are not a thing we think about much, if at all, in our daily lives. We are experiencing the effects, of course, but perceptions are that the consequences are not yet severe. Some mistakenly believe they are too far off into the future to take seriously in the present tense and that there is no certainty about the effects yet. This gives rise to the question of whether we are willing, or unwilling, to make present-day sacrifices for future generations. The answer may well depend on what we know or do not know about the threats not only to future generations but more directly to our future selves. Emissions reduction targets are harder to generate excitement about compared, say, to better flood control measures protecting your business or home from an anticipated or predicted increase in severe weather events. But we do have more than our faulty perceptions and our individual or localized experiences to guide our thinking. We have science.

The physical environment is constantly changing. This is true as a general observation. Being aware of this and designing our communities and living our lives so that they consider these changes, or act in harmony with them, if you will, is a practical and necessary thing. Today we know that the warming of the climate is projected to result in more dramatic meteorological events. Storms, floods, extreme temperatures, drought, wild fires, and all other natural hazards will be altered accordingly. These changes suggest that the nature and the course of those hazards that may be anticipated on a routine or regularly recurring basis in every community are subject to changes. The past is no longer a reliable source of information about the future, and risk and vulnerability profiles are changing. From an emergency management perspective, this adds up to a requirement that we be forward looking and anticipatory as never before as we integrate these changing climatological conditions and the science that explains them into our planning for the future. We must be prepared for the predictable impacts of these changes, and we must manage the risks they portend. This means we must be ready to adapt to changing conditions, and, where possible, mitigate climate change through the adoption of policies and strategies that may address its causes. We must not only adapt; we must mitigate.

A growing number of Americans do seem to support the concepts of climate adaptation and climate mitigation. But as we shall see in the next chapter, they have good reason not to feel very confident that federal, state, and local policies will ultimately help reduce the threats of global warming. They are also skeptical that federal, state, or local policies will help protect communities from the impacts of global warming. It is not so much that Americans believe that there are no policy initiatives that can make a difference as it is a lack of confidence that elected officials are willing and/or able to make the necessary effort. While there is a partisan divide in public opinion, most Americans would favor passing laws aimed at increasing energy efficiency and the use of renewable energy as a way to reduce our dependency on fossil fuels. Whatever their positions on the topic of climate change, a significant majority of Americans agree that developing sources of clean energy should be a high or very high priority for the president and Congress. But the policy environment in Washington, as well as in many state capitals, is somewhat inhospitable to initiatives that would enhance adaptation or promote significant reductions in fossil-fuel emissions. Partisan politics has produced a gridlock with respect to climate policy just as it has with respect to

other policy areas. But even if the policy environment were supportive in a bi-partisan fashion and nurturing of efforts to adapt to and
mitigate climate change, there is another problem. That problem is
the number of policies adopted and implemented without controversy or partisan disagreement that actually contribute to the net increase of greenhouse gas emissions.

As we have seen in earlier chapters, some U.S. politicians and business leaders refuse to acknowledge climate change as a serious problem. This is often because they have close dealings with fossil-fuel
companies and with states benefitting from employment in those industries. It is also undeniably true that the climate denial lobby has
successfully influenced the political process to delay adequate response to the climate crisis. But it is also true that we have seen consistent failures by American policy makers who do want to manage
climate-change hazards and yet seem wholly unable to do so.
Unfortunately, sincere but not sufficiently informed or committed climate policy makers may in practice be even more harmful than short-
sighted lobbyists and partisan sponsors of denial.

As we now turn our attention to the policy arena, we will see that
there is both potential for progress on the American political scene
and a concerted effort to impede that progress. For every opportunity to move forward, there is a huge obstacle to be overcome.
Moreover, for every positive step that is taken forward, it still seems
we are often taking two steps backward. My own analysis up to this
point convinces me that the ultimate task to be accomplished if
climate adaptation and climate mitigation are to be genuinely possible is nothing short of a transformation of the policy environment
itself. Whatever disagreements might persist in the policy dialogue,
that dialogue must begin with agreement on one central point. The
policy environment must accept that the climate crisis is real and
our need to deal with it is urgent. It must recognize what I have
maintained throughout our discussion. With respect to climate
change (a.k.a global warming), we have reached critical event,
or point of decision, which, if not handled in an appropriate and
timely manner (or if not handled at all), will turn into a disaster or
catastrophe.

References

1. Ravetz, J.R. (2006). "Post-Normal Science and the Complexity of
Transitions toward Sustainability." *Ecological Complexity* 3: 275–284.

2. Sadar, Z. (2010). "Welcome to Post-Normal Times." *Futures* 42(5): 435–444.

3. Saloranta, T.M. (2001). "Post-Normal Science and the Global Climate Change Issue." *Climate Change* 50: 395–404.

4. Britton, N.R. (2001). "A New Emergency Management for a New Millennium?" *Australian Journal of Emergency Management* 16(4): 44–54.

5. Beatley, T. (1995). "Planning and Sustainability: A New (Improved?) Paradigm." *Journal of Planning Literature* 9(4): 383–395.

6. Mileti, D.S. (1999). *Disasters by Design: A Reassessment of Natural Hazards in the United States.* Washington, DC: Joseph Henry Press, Environmental Studies.

7. Schneider, R.O. (2002). "Hazard Mitigation and Sustainable Community Development." *Disaster Prevention and Management* 11(2): 141–147.

8. Climate Change Adaptation in the Emergency Management and Critical Infrastructure Sectors—Workshop Proceedings (2011). http://www.cleanairpartnership.org/files/Emergency_Management_Workshop_Proceedings_Final.pdf (accessed August 27, 2015).

9. Labadie, J.R. (2011). "Emergency Managers Confront Climate Change." *Sustainability* 2011 (3): 1250–1264.

10. Silverman, J., Levy, L.A., Myrus, E., Koch, D.M., and DeGroot, A. (2010). "Why the Emergency Management Community Should be Concerned about Climate Change." *CNA Analysis Solutions.* Available at http://www.georgetownclimate.org/resources/why-the-emergency-management-community-should-be-concerned-about-climate-change-a-discussi (accessed August 28, 2015).

11. Intergovernmental Panel on Climate Change (2014). Fifth Assessment report. http://ipcc.ch/report/ar5/index.shtml (accessed July 3, 2014).

12. EPA (2014). Inventory of Greenhouse Gas Emissions and Sinks: 1990–2012. http://www.epa.gov/climatechange/ghgemissions/usinventoryreport.html (accessed July 1, 2014).

13. Howarth, R.W., Santoro, R., and Ingraffea, A. (2011). "Methane and the Green-House-Gas Footprint of Natural Gas from Shale Formations." *Climate Change* 106: 679–690.

14. Shindell, D.T., Faluvegi, G., Koch, D.M., Schmidt, G.A., Unger, N., and Bauer, S.S. (2009). "Improved Attribution of Climate Forcing to Emissions." *Science* 326: 716–718.

15. Jamarillo, P., Griffin, W.M., and Mathews, H.S. (2007). "Comparative Life-Cycle Emissions of Coal, Domestic Natural Gas, LNG, and SNG for Electricity Generation." *Science and Technology* 41: 6290–6296.

16. Jeans, H., Ogletorpe, J., Phillips, J., and Reid, H. (2014). "The Role of Ecosystems in Climate Change Adaptation." In *Community-Based Adaptation to Climate Change.* Schipper, E.L.F., Ayers, J., Reid, H., Huq, S., and Rahman, A., eds. London: Routledge.

17. National Climate Assessment (2014). http://nca2014.globalchange .gov/ (accessed July 3, 2014).

18. Giddens, A. (2009). *Politics of Climate Change*. London: Polity.

19. Klein, N. (2014). *This Changes Everything*. New York: Simon and Schuster.

20. City of Keene and ICLEI (2007). Keene, New Hampshire—Adapting to Climate Change: Planning a Climate Resilient Community (PDF). City of Keene, ICLEI http://www.ci.keene.nh.us/sites/default/files/Keene%20Report _ICLEI_FINAL_v2_0.pdf (accessed August 28, 2015).

21. USGCRP (2009). *Global Climate Change Impacts in the United States*. Karl, T.R., Melillo, J.M., and Peterson, T.C., eds. United States Global Change Research Program. New York: Cambridge University Press.

22. South Florida Regional Planning Council (2014). Climate Change Community Toolbox. http://www.sfrpc.com/climatechange.htm (accessed July 10, 2014).

23. Environmental Protection Agency (2014). State and Local Climate Impacts and Adaptation. http://www.epa.gov/statelocalclimate/local/topics /impacts-adaptation.html (accessed June 23, 2014).

24. Hawaii House of Representatives (2014). H.B. No. 1714. http:// www.capitol.hawaii.gov/session2014/bills/HB1714_.HTM (accessed June 23, 2014).

25. Galford, G.L., Hoogenboom, A., Carlson, S., Ford, S., Nash, J., Palchak, E., Pears, S., Underwood, K., and Baker, D.V., eds. (2014). *Considering Vermont's Future in a Changing Climate: The First Vermont Climate Assessment*. Gund Institute for Ecological Economics.

26. CA.GOV (2014). California Climate Adaptation Guide. http:// resources.ca.gov/climate_adaptation/local_government/adaptation_policy _guide.html (accessed June 23, 2014).

27. Chicagoclimateaction.org (2014). Chicago Climate Action Plan. http://www.chicagoclimateaction.org/pages/adaptation/11.php (accessed June 23, 2014).

28. Burton, I. (1996). "The Growth of Adaptation Capacity: Practice and Policy." In *Adapting to Climate Change: An International Perspective*, Joel B. Smith, ed. New York: Springer.

29. NOAA (2014). Adapting to Climate Change. http://coastalmanagement .noaa.gov/climate/docs/ch5adaptationstrategy.pdf (accessed June 30, 2014).

30. National Climate Assessment (2014). http://nca2014.globalchange .gov/ (accessed July 3, 2014).

31. Zimmerman, J. (2012). "North Carolina Tries to Outlaw Sea-Level Rise." *Grist* August 2, 2012. http://grist.org/list/north-carolina-tries-to -outlaw-sea-level-rise/ (accessed July 5, 2014).

32. Latin, A. (2012). *Climate Change Policy Failures: Why Conventional Mitigation Approaches Cannot Succeed*. Singapore: World Scientific Publishing Co. Pte. Ltd.

33. Fountain, H. (2015). "Panel Urges Research on Geoengineering as a Tool against Climate Change." *New York Times*, Feb. 10, 2015. http://www.nytimes.com/2015/02/11/science/panel-urges-more-research-on-geoengineering-as-a-tool-against-climate-change.html?_r=0 (accessed February 11, 2015).

34. Appell, D. (2013). "Strange Bedfellows? Climate Change Denial and Support for Geoengineering." Yale Climate Connections October 13, 2013. http://www.yaleclimateconnections.org/2013/10/strange-bedfellows-climate-change-denial-and-support-for-geoengineering/ (accessed August 27, 2015).

CHAPTER 6

Policy Options and Prospects

Introduction

In this chapter we will discuss and assess U.S. climate policy options from the national perspective. Any chance we will actually have to manage the climate crisis will be heavily dependent on what happens at the national policy level. At least as far back as the early 1990s, American presidents, starting with President George Herbert Walker Bush, have understood the need and have actually set the goal of reducing America's carbon emissions. They have failed to deliver. Emissions have continued to rise sharply, reaching their peak under President Obama. But emissions rates have finally begun to drop some in the last couple of years, and many expect them to continue to do so. By 2020, according to some sources, U.S. emissions rates will be 16 percent below 2005 levels. (1) This may be an optimistic projection. But even if it is accurate, it is not in and of itself nearly enough to even begin addressing the climate crisis.

When he took office in 2009, President Obama pledged that by 2020, the United States would achieve reduced greenhouse emissions rates of 17 percent from 2005 levels. With the continued failure of Congress to enact any comprehensive climate legislation, the prospects for meeting this goal seemed dubious. But despite congressional inaction, it seems we may come close. It appears that greenhouse gas regulations under the Clean Air Act are, with other factors, beginning to have an impact. The Clean Air Act, passed in 1970, gave the Environmental Protection Agency (EPA) authority to develop regulations that mitigate the harm of air pollution. This broad authority has been an important factor in the progress that has been made. In

2007, the U.S. Supreme Court ruled that the authority of the EPA to regulate pollution extended to the regulation of greenhouse gases. The EPA subsequently made a science-based determination that greenhouse gases are dangerous to human health and to the environment. Accordingly, the EPA has set about the task of regulating emissions.

More stringent vehicle fuel economy standards as mandated by the Energy Independence and Security Act of 2007 went into effect in 2011. The EPA, pursuant to its regulatory mission and authority, developed new rules in 2011 for the permitting of stationary structures such as power plants and industrial facilities and is beginning to regulate operating performance standards affecting new and, in particular, existing stationary facilities. The EPA is working to identify opportunities to reduce emissions from these sectors by up to 10 percent, or 6.2 percent of total U.S. emissions. This includes energy and process efficiency improvements, the beneficial use of process gases, and limited material and product changes that can be cost effective, meaning that they are zero-cost options for a firm after accounting for the cost of energy saved. (1) But before we allow ourselves to be too optimistic in light of these EPA actions, we might want to take a deep breath. Let us remember that nothing is really permanent here. The priorities are subject to change with a change in presidential administration. Additionally, the regulations we are talking about are fairly moderate. They are not aggressive enough to do more than reduce the rate of carbon and GHG emissions over the period of a few decades. This leads critics to suggest that we are delaying or back-loading important initiatives and efforts that are needed much sooner if we are to ultimately address the problem. A gradualist approach, they argue, allows the CO_2/GHG impacts to get much worse and much more expensive to address, as it spreads the effort to reduce emissions out over too long a time frame. Indeed, a gradualist approach may work to ensure that we will pass tipping points that are irreversible and will make it impossible for us to deal with the climate crisis, and thus, lead to a global disaster. We must also note that in response to the cautious but positive tweaks—and that is what these regulations being implemented by the EPA are—Congressional Republicans (who after the 2014 midterm elections now have majorities in both the House and the Senate) are eager to move aggressively to undermine the authority of the Environmental Protection Agency. Some even want to eliminate the agency altogether.

In addition to regulatory action by the EPA, other variables have contributed to the slight reduction in the rates of emission. Emissions have been slightly reduced by market trends such as a substantial shift toward the use of natural gas and away from coal for electricity generation. Over the long haul, however, there is reason to doubt that natural gas will have a positive impact on emissions. As noted in Chapter 5, the GHG footprint of natural gas is actually greater than we might think. It is also true that there is a discernible reduction in the energy intensity of economic activity stemming from the expanded role of energy efficiency. (1) But as noted in Chapter 5, it is equally important to note that the economic recession has also had an overall influence on reducing current emissions. This will have less of an influence by 2020. As the economy improves, emissions will probably rise again. Even were the progress made in the reduction of greenhouse gas emissions rates to be maintained or improved slightly, such emissions will continue to be a good deal higher than they would be if we actually passed and implemented comprehensive climate legislation. The prospects of that happening any time soon do not appear to be very great.

We can hope that the GHG emission trajectory of the United States will continue to improve, perhaps even accelerate. President Obama, in 2015, pledged to significantly increase reductions in U.S. carbon emissions in advance of the December global climate summit in Paris. But there are many indicators that this will require more effort than we seem to have the political will to undertake at the moment. Additionally, as we shall see in our discussion in this chapter, we continue to see nonclimate policies that are embraced by all partisans and enacted without much opposition that have the persistent effect of increasing net greenhouse gas emissions. For every step forward in the effort to reduce emissions, we do take one or two backward, or so it often seems.

When it comes to climate change mitigation, the reduction of greenhouse gas emissions is the only response that will mitigate or address the causes of global warming. The need to do this is logical if one considers the risks and vulnerabilities associated with a warming climate and their impacts on humanity and the environment. There is no escaping irreversible damage without addressing the excess of GHG accumulation in the atmosphere. But given that we will have some very negative impacts to deal with in each region of the country even if we get serious about mitigation—reducing emissions—we will also need to continue adapting to climate change. We have seen in our discussion in Chapter 5 that climate-adaptation efforts are underway across the nation. But these efforts are just beginning. Despite the modest steps being taken at the national level, and despite some impressive action at state and regional levels, we do need a national policy environment that is both more focused and inclusive. The coordination of efforts across jurisdictional lines requires national and international policy initiatives, as well as regional and local initiatives.

Thinking in terms of the United States, we urgently need the structure of policy and law to ensure a national commitment that recognizes the challenge of climate change as a crisis and the need to respond to it. This strikes me as a very logical statement. But it also occurs to me that this is something that is ill-suited for two-, four-, or six-year election cycles, and it is totally beyond the 24/7 news cycle's ability to understand or to convey. Speaking both logically and idealistically, public policy is what we need most. Public policy that recognizes and sets the priorities for action, public and private, can be considered the highest possible expression of a cross-party consensus recognizing the need to act on climate change. Institutionalizing this consensus is the best means of ensuring commitment, promoting timely action, assuring the continuity of action, providing certainty for investors, enhancing our credibility, engaging the public, and actually solving problems. It can be said, and not just in relation to climate change, that the United States is presently light years away from an ideal process that does any of these things that we might associate with a sane and healthy policy environment. Despite the ever more dysfunctional nature of our political and policy environment, there is some good news. We do know what to do and how to do it. Proposals for fruitful action exist. We have the knowledge needed to begin to address the climate crisis. It is the political leadership and the political will to act that is missing.

The major purpose of this chapter is to discuss the policy options that are being articulated but not acted upon, as we hope to respond to climate change by mitigating its causes and managing or adapting to its effects. We will discuss the policy initiatives that are available to choose from, the obstacles to adopting and implementing them, and the prospects for any real progress in climate mitigation and adaptation. We will conclude this discussion with a suggestion of what might be needed to make the policy environment a little more healthy and productive with respect to climate change.

Basic Policy Options

What are the policy options for action at the national level? What has our national government been discussing and considering as the impacts of a warming climate have become more apparent? The basic policy options for addressing the challenges of climate change fall into two categories. These are policies aimed at mitigation (addressing the causes) and policies aimed at adaptation (addressing the effects). It is very generally agreed that the reduction of carbon and other greenhouse gas emissions is a mitigating action that must be taken in response to climate change. The scientists have been saying this for almost 50 years (Chapter 3), and increasingly we have seen that governments around the world are in agreement. But the reduction of emissions cannot occur without a significant policy intervention. Why is this so? The marketplace is driven by short-term considerations, and its incentives do not always reward or encourage the optimal long-term choices for society. Often, as negative effects of economic activity are created and the costs of them are passed on to consumers, to society at large, to the future, and to the environment, there are no incentives that encourage the producers of these negative externalities to change their practices to reduce their costly and burdensome impacts. With respect to carbon emissions, the market fails to address the problem largely because there is no price or cost associated with carbon emissions that the producers must bear. But there is a cost to reducing these emissions. Producers will not incur these costs if they do not have to because they are driven by the short-term goal of profit. Their market incentive is to pass these costs off. Absent any incentives that make the carbon problem and the costs associated with it their problem and their costs, the producers will be content, incentivized actually, to ignore it. Anthropogenic climate change may thus be thought of as what economists call a "market failure."

The basis of the climate change market failure might be called the greenhouse gas externality. In other words, greenhouse gas emissions are a side effect of economic activities. The impacts of the greenhouse gas emissions do not fall on those who are conducting the activities that contribute to the problem. Instead, the impacts of these emissions fall on future generations, the environment, and people living in impacted areas. Those responsible for the emissions do not bear the cost for them. In this sense the negative effects of greenhouse gas emission are external to the market. The adverse effects of greenhouse gas emissions are not addressed by the market because there is in fact absolutely no market incentive to do so. As a result, the market continues to fail by overproducing greenhouse gases to the detriment of all.

An example of how market incentives often produce failures to manage negative externalities might be seen in the construction trade. Developers and builders have a short-term incentive to reduce the construction cost of the commercial structures and homes they build. The market influences the choices they make in a variety of ways. Reduction of construction costs increases profits. Reduction of construction costs keeps the initial purchase price lower or more competitive in the market. The market thus provides incentives to cut costs by skimping or cutting production costs on more efficient design features and systems that reduce energy consumption and that may reduce the carbon imprint. Since the cost of energy consumption is not borne by the builder but is passed off as an externality to the purchaser, and since the costs of the risks and vulnerabilities associated with the continued loading of carbon into the atmosphere are not borne by the builder but are passed off to society at large, the market fails to incentivize choices that are optimal for carbon reduction and of greater social value. The concept of incentivized market failures, as exemplified in this construction example, in relation to the reduction of carbon emissions is the rationale that is used to justify policy intervention as a means of addressing the problem. Economists might say, as most do in fact, that there is a need for a policy intervention to increase the price of activities that emit greenhouse gases as a means to guide market decisions and to stimulate innovation in the form of low-carbon technologies. To ensure that emissions are in fact cut and that the costs of this reduction are spread out across the economy as inexpensively as possible, economists generally favor policy interventions that require all businesses and households to share equally a price for carbon.

There are three basic policy options that are most frequently discussed in the United States with respect to climate change mitigation. Remember, *mitigation* means reducing carbon emissions. These three policy options are a carbon tax that requires producers and consumers to bear some of the costs, a cap-and-trade system that restricts the amount of carbon that can be produced and that allocates emissions through a market-based trading system, and stricter and direct regulation of carbon emissions. The first two options, a carbon tax and a cap-and-trade system, seem to be the preferred options. This is based on the assumption that the most efficient approach for emissions reductions is to give businesses and households an economic incentive. It has become an accepted orthodoxy in the United States that the most efficient approaches to reducing emissions involve giving businesses and individuals an incentive to curb activities that produce CO_2 emissions rather than adopting a command and control or regulatory approach in which the government would mandate how much individual entities could emit or what technologies they should use. Let us briefly examine the two preferred options.

A 2008 Congressional Budget Office study concluded that a tax on carbon emissions would be the most efficient incentive-based option for reducing emissions. (2) A carbon tax is, in essence, a pollution tax. It imposes a fee on the production, distribution, and use of fossil fuels based on how much carbon their combustion emits. The government sets the rate per ton for carbon emissions and then translates this into a tax on the production and consumption of things like electricity, natural gas, and oil. The basic assumption behind this tax is that it will make using dirty fuels more expensive. This in turn will encourage utilities, businesses, and individuals to reduce consumption and increase energy efficiency. A carbon tax is also thought to be a means to make alternative energy more cost-competitive with cheaper and polluting fuels like coal, natural gas, and oil. According to the Congressional Budget Office, a very modest carbon tax of 25 dollars per metric ton of carbon would, over a 10-year period, reduce emissions by 10 percent and generate a trillion dollars in new federal revenues. (2)

As presented by its proponents, a carbon tax is the most direct and the most efficient way of offsetting the negative externalities of the market. Externalities, once again, are costs or benefits generated by production of goods and services. Negative externalities are costs that are not paid for or costs or harms that are passed on to others. When utilities, businesses, or homeowners consume fossil fuels, they create pollution that has a societal cost. Everyone suffers from the effects of this pollution. Proponents of a carbon tax believe that the price of fossil fuels should account for the societal costs they create. In other words, if you're polluting to everyone else's detriment, you should have to pay for it. In addition to discouraging fossil-fuel use and lowering the rates of carbon emission, the federal revenues generated by a carbon tax could be used to support and incentivize the production of alternative clean energy options. While a carbon tax has considerable support and is a preferred option among scientists and among economists across the political spectrum, it remains politically impossible to adopt and implement it in the United States. I will elaborate on the reasons for this later, but first let us continue our review of the basic policy options. Next, we will discuss the policy option known as cap and trade.

Proponents of cap and trade as a policy response to the market failures that create too many carbon emissions believe that it can deliver significant results with a mandatory cap on emissions and that it can do this without inhibiting economic growth. It is typically said that cap and trade provides polluters flexibility in how they will comply, rewards innovation and efficiency, and achieves significant additional emission reductions with minor inconvenience for the polluter. In a cap-and-trade system, the government sets maximum allowable emissions by law. This is the cap, and it applies to all polluting industries. For every ton of CO_2 a polluter, a power plant for example, reduces under the cap that has been set for it, it is awarded one allowance. As an entity accumulates allowances, they can be sold, traded, or banked for the future. Any polluter that has successfully reduced emissions below their mandated level can auction off or sell their allowances to those who are over-polluting. This, it is said, is a built-in cash incentive to reduce emissions that encourages compliance, promotes innovation, and enables the market to efficiently reduce emissions. Well, that's the theory.

A cap-and-trade system is popular with those who believe that a carbon tax is punitive and/or with those who believe, just because it is a tax, it is impossible to enact. There is also the notion that a cap-and-trade system more responsive to market changes will generate less political opposition than a carbon tax. In fact, this belief explains why every serious proposal that has been put forth in Congress (and of course rejected) in recent years has featured cap and trade as the preferred mechanism for using the market to reduce carbon emissions. The United States actually has had success with a cap-and-trade policy approach in the past. This was in connection with the Acid Rain Program. The stated objective of this EPA program was to "achieve significant environmental and public health benefits through reductions in emissions of sulfur dioxide and nitrogen oxides." (3) In essence, this program sought to reduce the damage caused by acid rain. It did this by creating a cap-and-trade system for sulfur dioxide and nitrogen oxide emissions. The program has been successful and has led to improvements in air quality, acid deposition, and surface water chemistry. This success has convinced many that a similar approach can work with respect to the goal of reducing carbon emissions. Despite its relative popularity as a policy option, a cap-and-trade policy has been every bit as impossible to adopt and implement as a carbon tax. Again, the reasons for this will be discussed later. But let us continue our look at the options.

As we have already noted, option number three, the direct regulation of carbon emissions, has been recently implemented and with some success. Regulatory action by the Environmental Protection Agency (EPA) has helped to reduce emissions rates. More stringent vehicle fuel economy standards, new permitting requirements for stationary facilities, and increased performance standards are making a contribution. But this progress is offset by the reality that the regulatory steps taken are relative baby steps when it comes to actually addressing the GHG problem. One must also take into account the increasing hostility to regulation in general and environmental regulation in particular. Most of what we might call the orthodox American thinking about the matter accepts that regulation alone, as presently and cautiously implemented, is not sufficient to reduce carbon emissions at a fast enough rate to really address the problem. The argument is also made and generally accepted that a tougher regulatory approach would be politically unacceptable. Most suggest that the best policy approach is one that combines modest but more widely acceptable government regulation, a carbon tax, and a cap-and-trade system. The conventional wisdom here is that a regulatory approach must of necessity be moderate to be acceptable or politically feasible and that it will be most effective in concert with the other two market options (carbon tax and a cap and trade) in addressing the problem over time. But as climate change is loading the dice toward more extreme hot weather and the impacts that we are already feeling appear to be escalating, some question the efficacy of these three favored policy approaches. Other climate change mitigation options, more dramatic and less popular perhaps, must be noted in this context.

It must be remembered that the major factor determining the extent and seriousness of global warming is the cumulative atmospheric concentration of carbon and other greenhouse gases. It is not only the rate of annual emissions that must be addressed but the cumulative effect. In fact, it might be suggested that the gradual reduction of emissions rates over a period of many decades (which is what carbon taxes, cap-and-trade systems, and moderate regulation all aim at) is an ill-advised way to address the problem because it continues to allow the accumulation of more greenhouse gases in the atmosphere. This gradualist approach, it might be said, may slow the rate of growth in greenhouse gas emissions. But it might have minimal value with respect to reducing climate-change-related risks and vulnerabilities. In other words, the gradual approach creates the illusion of climate change mitigation, but its net result may be little more than a waste of irreplaceable time and resources that could have been expended in more fruitful efforts. It postpones necessary action and runs the very real risk of being too little too late. The result is the back-loading to later decades of too many important steps that need to be taken right now. This point of view suggests that we need to move beyond policies that gradually reduce emissions to a more aggressive approach that will decarbonize our energy economies.

For the United States to remove carbon emissions from its energy system in, say, 20 years, it would have to build new clean energy capacity at 10 times the rate possible in its built or existing capacity. Decarbonizing our energy economy is a big (some might say impossible) challenge. But there is no doubting the need to develop replacement and clean energy alternatives as quickly as possible. This is a stated goal of the Obama administration, for example, and it is a goal widely appreciated and shared. We should have started to prioritize clean alternative energy development decades ago, and there is no doubt that the odds of ever decarbonizing our energy systems become longer with each passing day. A shift to cleaner and renewable energy sources is perhaps the number-one policy priority for those who feel that gradually reducing emissions is simply too little too late. The argument to be made in this context is that our policy priorities should be to stabilize the accumulated greenhouse gas concentration as low as possible and develop replacement technologies as quickly as possible. This means we must stop adding to the problem. This means we must ultimately shut off the fossil-fuel faucet entirely.

All of the policy options discussed up to this point have to do with mitigation or addressing the causes of global warming. Whether we are talking about efforts to reduce carbon emissions gradually, to spread the costs of carbon equally, or to be aggressive about decarbonizing the atmosphere through the development of alternative energy sources, we are talking about the causes (i.e., greenhouse gas emissions). But as we have seen in our discussion of risks and vulnerabilities in Chapter 4, climate change is already affecting communities, livelihoods, and the environment across the United States and around the world. As we have seen in Chapter 5, climate-change adaptation is increasingly being integrated into planning and policy initiatives at the local and state levels. It is also a priority in federal agencies. Adaptation to global warming is, in essence, responding to the effects of climate change. Adaptation policy needs to be anticipatory. This means it must be based on the projected risks and vulnerabilities that must be managed in relation to climate-change impacts. As we have seen in Chapter 5, adaptation policy includes things such as policy options for the adaptation of infrastructure, agriculture, coastal areas, and water resources to current and future impacts. Most of the adaptation activity takes place at the state and local level, but federal support can be a critical component in promoting and supporting this activity.

In November of 2013, President Obama issued an executive order on preparing the United States for the Impacts of Climate Change. (4) This executive order directed federal agencies to take a series of steps to make it easier for American communities to strengthen their resilience to extreme weather and prepare for the other impacts of climate change. It instructed agencies to update all federal programs to support climate-resilient investments and to plan for climate change related risks to federal facilities, operations, and programs. It also instructed federal agencies to provide the information, data, and tools that state, local, and private-sector leaders need to make smart decisions to improve preparedness and resilience. (4)

In summary, the basic policy options that are most widely considered as viable within our national political dialogue (though not all—few in fact—have been enacted) include the carbon tax, a cap-and-trade system, regulatory initiatives to reduce carbon emissions by improving fuel standards, promotion of alternative renewable energy development, and climate preparedness or adaptation policies aimed at protecting and enhancing hazard resilience in the face of predicted impacts. As we turn to an assessment of exactly what has transpired on the policy front, it becomes apparent that despite bits and pieces of small progress, climate-policy development has had more failures than successes, especially in recent years. A brief explanation of the reasons for this may give us reason to be less than optimistic about the future of climate policy as well. The United States, for all of its wealth and all of its technological capabilities, is proving to be less than adequate in meeting the challenges of the climate crisis.

U.S. Climate Policy Initiatives, Prospects, and Obstacles

In the summer of 2013, the White House unveiled President Obama's U.S. Climate Action Plan. (5) This plan consists of a wide variety of executive actions designed to achieve three major goals. Goal number one is to cut carbon pollution in the United States. But instead of legislative proposals (e.g., a carbon tax, a cap-and-trade regime, new regulations), the plan says that the president will act through executive orders and the existing regulatory authority of federal agencies (EPA) to put tough new rules in place to cut carbon pollution. Goal number two is to prepare the United States for the impacts of climate change. With respect to the anticipated impacts of climate change, the president articulates the priority of helping state and local governments strengthen their roads, bridges, and shorelines so they can better protect people's homes, businesses, and ways of life from severe weather. But aside from a highway bill introduced in the summer of 2014 and hopelessly stalled in Congress, no specific legislative proposals have followed. The third major goal is for the United States to lead international efforts to combat climate change and prepare for its impacts. (5) What this means in practice remains to be seen. In reality, the focus of the Climate Action Plan is on what can already be done, and is in many instances being done, under existing executive authority. Beyond that, the 2013 plan was little more than a wish list that laid out what might be called a foundation or a very general road map for longer-term and very vaguely specified future policy actions.

A legitimate criticism of the Climate Action Plan from the perspective of environmentalists and climate science is that it is too short on substance and much too exaggerated in rhetoric. In the view of some critics, the actual progress it will enable us to make in responding to the climate challenge will amount to little more than a pinky in a crumbling dike if carbon dioxide continues building up at anything like the rates we have seen and that we anticipate. But before being too critical of the president's plan, and there is much to criticize, one must consider the policy environment in which it was developed. Congress has never produced anything close to a comprehensive climate and energy bill. It has never passed a significant bill to reduce carbon emissions. The closest it has come in recent years was the Waxman-Markey carbon regulation bill in 2009, as discussed in Chapter 3. This legislation included a cap-and-trade provision, and it passed the then Democratic-controlled House by a vote of 219–212.

It was unable to pass the Senate, however, where it needed a filibuster-proof 60 votes to pass. This was unattainable. With Republican control of both chambers of Congress firmly entrenched after the 2014 midterm election, the prospect for any meaningful legislation simply has not existed since that time. It is no mystery why this is the case.

Republicans are united and uncompromising in opposition to any climate-policy initiatives. They are as much opposed to President Obama's agenda as to any specific legislation. Getting any climate or clean energy initiatives passed is more than unlikely; it is impossible. Consider, for example, that the Republican Party Platform of 2012 stated the party's opposition to "any and all cap-and-trade legislation." This in spite of the fact that cap and trade was a market approach originally designed by and popular with Republicans. In addition, the platform demanded that Congress "take quick action to prohibit the EPA from moving forward with new GHG regulations." (6) On every front, it appears that the Republican Party is united and unshakeable in its opposition to action in response to climate change. In fairness, Republicans are frequently aided in opposition to climate-policy initiatives by Democrats from coal-producing or fossil-fuel-producing states as well.

The fact of the matter is that congressional action will be required if the United States is to make any progress in responding to the climate crisis. This is not only necessary for the adoption and implementation of measures where there is a growing consensus (e.g., carbon tax, cap and trade, promotion of alternative energies) that action can make a positive difference, but also with respect to existing federal laws adopted before climate change was a consideration (e.g., the Clean Air Act) and that must now be adapted to regulate greenhouse gases. (7) But the prospects for any meaningful congressional action simply do not exist on the immediate horizon. Given the dynamics of the policy environment at the beginning of his second term and the unlikelihood of any legislative action, President Obama had no recourse but to unveil his Climate Action Plan and what, absent any legislative action, can only be a half-baked program to address climate change primarily through already established executive regulatory action. It should also be noted that, its limitations understood, many believe the plan will make a positive contribution. (7) But the key point is, at least in the eyes of his supporters, there was not much else the president could do under the circumstances. Taking the initiative to act where he could, outlining (however vaguely) broad objectives for the future, and influencing the public discourse were the only options left to him.

Backing up and taking a longer view, and based on the science and what we honestly know, the entire policy dialogue seems out of step with reality on several levels. While it is true that new proposals to mitigate and adapt to climate change seemingly abound, these proposals may not be nearly enough to make a dent in the problem. As we have seen, the U.S. Congress has been unable to legislate, and the executive branch is acting where it can. On a state and local level, more and more efforts are in fact being aimed at what logically needs to be done in order to adapt to the risks and vulnerabilities of a warming climate. Mitigation has remained difficult to implement. Since the beginning of the 21st century, it can be said that governments around the world have steadily increased the number of climate laws adopted and enacted. The six largest global emitters of CO_2 have at least attempted to reduce emissions. It can be argued that the United States, the European Union, Japan, India, and China are actually making some small but measurable progress. Only Russia seems to be making no real progress on climate change mitigation efforts. (8) But despite all of this "progress," there is less forward momentum than we might like to think. In the United States especially, there are a number of factors that work against progress and forward movement.

We need to consider three extremely important and debilitating factors that offset, perhaps even overwhelm, the progress we think we are making. Each of these factors is a critical obstacle to progress, and they are especially observable in the American policy dialogue. First, there are too many of what we might call anticlimate policies that continue to make the problem worse faster than we can respond to it. Second, most of the policies we in the United States and countries around the globe are proposing, in some cases adopting, are deferring the actual reduction of carbon and other greenhouse gases to a later date or back-loading the bulk of the effort onto future generations. In other words, we are attempting to slow the annual rates at which we are emitting greenhouse gases, but we continue to increase their accumulation when we should have a greater sense of urgency about stabilizing the accumulated greenhouse gas concentration at the lowest level possible. Third, the politics of climate denial have become institutionalized and are deeply embedded in the American policy debate. Let's discuss each of these obstacles to progress in a bit more detail.

At the same time new climate policies are being debated, and, as there is a growing sense that we must reduce greenhouse gas emissions as a critical necessity, our government (and governments around the world) is introducing and implementing policies that have the opposite effect. (8) Permits are being issued at increasing rates for new offshore drilling sites and new coal mining ventures, and new relatively unregulated natural gas initiatives are expanding at warp speed. This is a juxtaposition that is more than just a bit jarring. As we are seeking to make progress with renewable energy and energy efficiency, as we are issuing a few new rules to limit carbon output, we continue to promote and boast about new record levels of carbon-fuel production. Indeed, American policy makers, including and especially many Democrats who sincerely want to address the threats associated with climate change, are absolutely thrilled that the United States is now the largest producer of natural gas in the world and is perhaps soon to become the largest producer of gas and oil in the world. As new warnings from the IPCC and the scientific community make us more clearly aware than ever before that the threats to life, health, and commerce posed by carbon emissions are growing, we pursue and boast about policies that contribute to record levels of carbon-fuel production. The science says, in effect, the priority should be to leave as much carbon (i.e., fossil fuels) in the ground as possible, yet the Obama Administration and American policy makers generally are in agreement across party lines that we must proceed with and accelerate activities (e.g., natural gas fracking, drilling for oil in the Arctic, new coal production leases and export terminals, etc.) that will ultimately render large parts of the planet uninhabitable and that will visit the worst impacts of climate change on all of us in the future. The contradiction between sincere but very meager climate initiatives and the aggressive heavy-handed policies we support that make the problem worse is very jarring indeed. Instead of pride about new record levels of fossil-fuel production, we should feel concern and perhaps even a bit of moral embarrassment. Our full-speed-ahead commitment to more and more extraction is, on the face of it, insane. Accelerating and enhancing the very activities that have caused the climate crisis will not solve it.

What, from a policy perspective, is really causing the greatest harm to us right now when it comes to the future of our climate policy? It is arguably energy subsidies. Instead of discouraging fossil-fuel use, the federal government underwrites it with tax incentives for the producers. Governmental subsidies for fossil-fuel producers reduce the cost to consumers for oil, natural gas, and coal. They have the effect of promoting wasteful use of the polluting and finite resources that are directly linked to anthropogenic or human-caused global warming. Tax subsidies have long been a major vehicle for U.S. energy policy. In addition, there are a number of nontax subsidies, such as the below-market leasing of federal lands for oil development, that along with tax subsidies provide billions of dollars in incentives for the producers of fossil fuels to keep on doing pretty much what they do. It is true that some subsidies go to renewable fuels, but these are next to nothing compared to what the fossil-fuel producers receive. Between 1994 and 2009, for example, the U.S. oil and gas industries received a cumulative $446.96 billion in federal subsidies. This compared to just $5.93 billion given to renewables in those years. (9) Subsidies for fossil-fuel producers may be the most difficult of the anticlimate policies to address. Despite the fact that they foolishly incentivize the very actions that will make the climate crisis more severe and unmanageable, they have also proven to be nearly impossible to eliminate. They are popular in fossil-fuel-production states, and their continuation is pragmatically seen as a political necessity for politicians seeking reelection.

Government subsidies effectively transfer a portion of the costs of production to taxpayers, enabling artificially low prices and inflated profits, while at the same time passing all of the costs of carbon pollution on to future generations and to the environment generally. This is a pretty good gig for the fossil-fuel producers. Their costs for the development and production of fossil-fuel energy have been and continue to be underwritten with our tax dollars to the tune of billions of dollars per year. The continued influence of fossil-fuel producers over governmental policy makers not only keeps the subsidies coming but also limits the advancement of policies that would make them pay a price for the carbon they produce (carbon tax or cap and trade), and the subsidies are so heavily weighted in their favor that they have the additional benefit of dampening competition from a competitively disadvantaged and weaker renewable energy market. (9)

Spending billions of dollars annually on policies that undermine our efforts to address climate change and that worsen the environment generally is neither logical nor profitable for us in the long run. But such policies will continue to be promoted. Fossil-fuel producers will continue to use their considerable influence to make this so. Politicians of both parties will be afraid to break the ties with this policy because the pervasive role of fossil fuels in the nation's economy and the influence of the traditional energy industries will continue to make it attractive and profitable or necessary (politically) for them to pursue anticlimate options. This in spite of the fact that a growing number of studies show that these subsidies are an incredible, almost total waste of money that would be better and much more productively spent on things like education and addressing poverty. (10)

To its credit, the Obama administration has maintained that climate policy is a critical priority for America's future and for our shared human existence in a warming world. But the anticlimate priorities supported by the president and others are as dangerous to our future as anything the climate deniers have to offer. In promoting an all-of-the-above energy policy approach that incorporates promoting renewable alternatives with an unregulated expansion in hydraulic gas fracking and an acceleration of offshore oil and gas development, the president is not only working at cross purposes but is sustaining a contradiction that is not in the end sustainable. You simply cannot continue to add greenhouse gases into the atmosphere and expect to address climate change. The tub is full and overflowing, and reducing the rate of the flow is not at all a reasonable response because we are still loading carbon into an already dangerously oversaturated atmosphere. At some point, we must turn the carbon faucet off. At some point, the damage does become too big to repair and the house (planet) is totaled and uninhabitable.

In addition to policies that work at cross-purposes, anticlimate policies that may erase any of the advantages of our climate initiatives, there is an additional and more basic obstacle to be overcome. As we have noted, most of the policies we in the United States are proposing, debating, and initiating in some cases are not really addressing the problem of greenhouse gas emissions directly. They are instead deferring the reduction of carbon to a later date or backloading it onto future generations. It is important to understand why this is a problem. When we hear that the United States has seen, due to some modest regulation and the economic downturn, a decline in the rate of carbon emissions and that by 2020 U.S. emissions will be 16 percent below 2005 levels, there may be a temptation to be optimistic. Despite inaction by policy makers, regulatory initiatives do seem to be making a difference. But we may be seduced by the illusion of tremendous progress if we are not fully informed. While even such small progress is good, it is still very small indeed. We really do need to understand that.

Consider that in 2012 the United States emitted a total of 14.4 trillion pounds (6,526 metric tons) of greenhouse gases into the atmosphere. The 2012 total represented a five percent increase in emission rates since 1990. It also represented a 10 percent decrease in emission rates since 2005. The things that determine the rate at which greenhouse gases are emitted include economic activity, energy prices, consumption patterns, land use, population, and technology. According to the EPA, between 1990 and 2012, the increase in greenhouse gas emissions corresponded with increased energy use by an expanding economy and population and an overall growth in emissions from electricity generation. The basic reality is, slight reductions in the *rate* of emission notwithstanding, we are continuing to add more greenhouse gases to the atmosphere when we should be trying to reduce the cumulative amount. As our discussion of the science in Chapter 2 revealed, CO_2 concentrations have continued to accumulate even with slight reductions in the rate of emission. CO_2 concentrations reached 400 parts per million (ppm) in 2013. March of 2015 saw 400-plus ppm concentrations hold uninterrupted for the full month. This was a first in human history. The scientific consensus is that we need to reduce CO_2 concentration to 350 ppm if we are to have any chance of avoiding the worst possible climate change scenarios. But even with progress being made in reducing emissions rates, it is highly unlikely that we will stop adding to the problem any time soon. More precisely perhaps, we will not make enough progress as soon and as completely as we need to. By the time we get to where it is we intend to go with our current policies, we will have arrived too late to achieve their purpose.

The approach of gradually reducing the rate of emissions does not in the end accomplish very much in relation to climate change. It is the cumulative amount that matters, and it continues to grow. We can say we are making some progress in reducing the rate of emissions, but we continue to add to CO_2 and greenhouse gas concentrations and thereby increase the negative and dangerous impacts of a warming climate. In this context, gradualist policy initiatives or market solutions that take decades to have an effect, like a carbon tax or cap and trade, and even some increased use of existing regulatory power, may be considered too little too late. Be that as it may, there is little likelihood that policies that would be more aggressive in attacking the problem would be feasible in the American policy process when we have not been able to achieve any consistent progress in adopting and implementing more generally popular but tepid measures.

As the United States continues to be lacking in either the ability or the political will to legislate gradual measures that are framed in the gentle language of market solutions (carbon tax or cap and trade) that will theoretically spread the costs of carbon fairly without damaging the economy or causing major inconveniences for anyone, the evidence may already be mounting that we are well past the point where our responses to the challenges of climate change can be a planned and gradual transition. The post-normal climate is already, in the eyes of many, producing profound and unwanted impacts that make a mockery of both our current policies and our debates about them. One need not be ready to accept or propose any specific remedy to concede that what we are presently doing may not be nearly enough. At a minimum, we can say it is becoming apparent that we might have to stop pretending that we can take the necessary steps to address the challenge of climate change without pain. There will be costs for businesses, households, and the economy to bear that will not be welcomed but that must be paid in one way or another. The only other alternative is unthinkable and much more expensive. Yet, as our continued bi-partisan support for contradictory anticlimate policies and the inability of our policy makers to enact even gradualist policies to deal with the causes of climate change reveal all too clearly, we are not nearly ready to have a conversation about bearing the costs and burdens that will only be made greater by our refusal to discuss them. Interestingly enough, the costs may not really be as great as some think if we move quickly enough.

The 2014 IPCC assessment report emphasized the need to take immediate and extreme steps to reduce the accumulation of greenhouse gases. These steps were said to include a carbon tax, cap-and-trade schemes, and more aggressive regulation together with a much more urgent effort to develop cleaner energy alternatives. Among the usual excuses for not doing so is the notion that such measures would be damaging to the economy. In other words, the mind-set that is promoted by those who oppose aggressive action is that you can have economic development that reduces poverty and assures the continued comfort of the well-to-do or you can combat carbon emissions, but you cannot do both at the same time. According to the IPCC calculations, the rate of economic growth ranges from 1.6 percent to 3 percent annually. The implementation of the measures it sees as critically important to address the excess of carbon in the atmosphere would reduce the annual growth rate by only 0.06 percent. Thus, the IPCC maintains that the economic argument against climate change mitigation is a false one. (11) But accurate or not, such calculations have little impact on American policy makers. This leads us to the third and perhaps most deadly obstacle to anything resembling climate policy in the United States. Even the most gradualist and least intrusive options, options where there is a growing public consensus or that rely on the ever-popular market, are dead in the water. Politics in the United States presently means no policy action with respect to the climate challenge.

As far as can be determined, there is only one political party in a democratic system that has denied climate science and that is nearly absolutist in its rejection of policies to address it. That would be the Republican Party in the United States. This is not simply a conservative point of view. For example, the Tories in Great Britain and Angela Merkel's Christian Democratic Union in Germany are conservative parties that have not denied the science associated with climate change. They have even taken policy initiatives to address it. As we saw in our discussion in Chapter 3, climate change has become a very partisan issue in the United States. We have seen that as the evidence about global warming has gotten stronger, the Republican denial of that evidence has only gotten more absolute. Understanding why this is so and why it effectively prevents meaningful policy action is critical if we are to ever make any progress at all. Understanding that our politics makes managing the climate crisis impossible is something we all need to consider and address if we think there is a crisis to be managed.

In January of 2014, the House Energy Committee once again voted down an amendment that would have simply stated that climate change was occurring. This amendment was proposed in connection with the Electricity Security and Affordability Act that sought to restrict or put an end to EPA regulation of emissions for new power plants. The bill made it through committee, but the amendment was voted down with all 24 Republican members voting against it. This was not the first time Republicans in Congress had unanimously rejected amendments stating that climate change was a reality. It won't be the last. Or will it? An interesting twist in the denial strategy occurred in January of 2015 when the U.S. Senate passed by a vote of 98–1 a resolution stating that climate change was real. Republicans voting for this resolution said the climate was changing and that it has always changed in the natural course of events, but that none of this is human caused. Thus, they could support the resolution yet still deny human-caused climate change. (12) In other words, climate change is real. The hoax is that it is human-caused and that it poses a serious threat requiring urgent action. It is almost as if denial of the science regarding anthropogenic climate change has become an enforced orthodoxy in the Republican Party. The reasons for this are a combination of factors that make it impossible for a Republican to be nominated for or elected to public office unless they are a climate-change denier. Let's speculate why this might be so. As we do so, remember our discussions and the information presented in Chapter 3.

Campaign finance law was revolutionized by the Supreme Court's decision in the famous *Citizens United* case. This ruling, notable for its determination that corporations were people possessing the right of freedom of speech under the First Amendment, led to a conclusion that money was speech. In other words, money contributed to political campaigns was entitled to full First Amendment protection and could not be limited. This, in turn, was interpreted to mean that individuals could give unlimited amounts of money to support candidates or causes as long as the money was controlled by an independent political action committee and not directly by a campaign or party organization. This opened up a floodgate of money from extremely wealthy individuals with an agenda. Whether in direct support for a candidate or in support of efforts by organizations that theoretically refrain from direct advocacy for candidates, the goal is clearly to push the public agenda in one direction or another. As we saw in Chapter 3, the climate denial movement is amply funded by invisible billionaires who can give unlimited amounts to candidates and causes of their choosing, and they can do so in relative secrecy. Anticlimate change activists like the Koch brothers have become the gatekeepers in Republican politics. Climate change denial is a price of admission for any Republican office seeker wanting access to the big money donors who support Republican candidates and causes. The Kochs have even gone so far as to create a pledge for officeholders and candidates to sign. This pledge is a promise to never support any legislation related to climate change that increases government revenues. The expectation is that candidates will also oppose and seek to prevent any EPA regulatory activities aimed at reducing GHG emissions.

To the influence of big money from wealthy individuals and opportunistic and well-funded climate-denial organizations, and the fact that access to this money is the lifeblood of Republican politics, can be added the increasing extremism and absolutism of the Republican base in the electorate. As noted in Chapter 3, political partisanship is the variable that most explains the variations in public attitudes about climate change. It is a well-established fact that voters in primary elections are among the most ideologically extreme in both political parties. Democratic primary voters are generally much more liberal than the general population, and Republican primary voters are much more conservative than the general population. The Republican base, the most conservative segment of the voting population, has become much more extreme in its anticlimate change views and much more absolutist in general with respect to its rejection of any governmental action to address the issue. In order to win a primary election, or if an incumbent wishes to fend off any primary challenge, climate change denial has become a necessary survival tactic. As the money and the party base reinforce Republican denial, they also contribute to the election of people who actually are deniers, and in some cases, so total in their denial that it is hard to imagine how anyone could make such preposterous and embarrassing statements and not actually believe them (see the following list). In other words, for an increasing number of elected Republicans, climate change denial is not an act or a fundraising tactic; it is an article of their faith. Many of them actually believe the unscientific and very foolish things they are saying. Almost every comment they make runs in the face of science and denies reality. The Republican dialogue about climate change covers familiar ground. It says climate change is not happening. If it is happening, it is natural and not human made. The science is not clear, or it is no big problem, and of course anything we do to address it will absolutely destroy the economy and our way of life. They—and we, by virtue of their successful efforts—are caught in a temporal time loop of nonsense that distorts our forward political movement and numbingly repeats itself without interruption and without our being entirely aware that all of this has been heard and rejected before. It is déjà vu all over again, as Yogi Berra might put it. Whether deniers as a matter of political expedience or true believers in the denial cause, Republican candidates for national office and elected office holders read from the same script.

Elected Republicans Speak on Climate

1. "Just so you'll know, global warming is a total fraud and it's being designed because what you've got is you've got liberals who get elected at the local level want the state government to do the work and let them make the decisions. Then, at the state

level, they want the federal government to do it. And at the federal government, they want to create global government to control all of our lives. That's what the game plan is."—Rep. Dana Rohrabacher, R-California

2. "This whole Al Gore thing of climate change unfortunately is not doing this nation any good."—Rep. Jeff Miller, R-Florida

3. "The government can't change the weather. . . . We can pass a bunch of laws that will destroy our economy, but it isn't going to change the weather. Because, for example, there are other countries that are polluting in the atmosphere much greater than we are at this point; China, India, all these countries are still growing. They're not going to stop doing what they're doing. America is a country, it's not a planet. So we can pass a bunch of laws that will do nothing to change the climate or the weather but will devastate our economy."—Senator Marco Rubio, R-Florida

4. "I don't agree with the notion that some are putting out there—including scientists—that somehow, there are actions we can take today that would actually have an impact on what's happening in our climate. Our climate is always changing. And what they have chosen to do is take a handful of decades of research, and say that this is now evidence of a longer-term trend that's directly and almost solely attributable to man-made activity."—Senator Marco Rubio, R-Florida—again

5. "I absolutely do not believe that the science of man-caused climate change is proven. Not by any stretch of the imagination. I think it's far more likely that it's just sunspot activity or something just in the geologic eons of time where we have changes in the climate."—Senator Ron Johnson, R-Wisconsin

6. "There isn't any real science to say we are altering the climate path of the earth."—Senator Roy Blunt, R-Missouri

7. "We came very close to adopting a cap-and-tax scheme that would have devastated our economy without a single demonstrable benefit. Now EPA has adopted greenhouse gas regulations on the basis of scientific assumptions that have been totally undermined by the latest science—and those regulations are going to have a devastating impact on many American families and businesses if we don't roll them back."—Senator Ted Cruz, R-Texas

8. "Global warming is 'unequivocal'? It's just flat not true!"—Rep. Joe Barton, R-Texas

Whatever the reasons that seem to compel Republican candidates and officeholders to be climate change deniers, and whatever our own views on the subject, it seems logical to conclude that with a Republican majority in the U.S. House, a Republican majority in the U.S. Senate after the 2014 midterm election, and given the degree and intensity of denial in the partisan ranks, no meaningful policy will be adopted and implemented any time soon to address the causes of global warming. This being the case, President Obama in 2014 and 2015 seemed to be trying to use his executive initiative yet again. In an agreement with the Chinese government, President Obama announced that he had agreed that the United States will cut net greenhouse gas emissions at least 26 percent by 2025. This would require doubling the current rate of U.S. carbon reduction. Chinese President Xi Jinping announced a stepped-up time frame for carbon reduction and for increasing China's use of nuclear, wind, and solar energy. (13) Most of the U.S. emissions reduction goal would be met presumably through existing authority and regulatory actions by the Environmental Protection Agency, the U.S. Energy, Interior, and Agriculture departments to target carbon pollution from power plants, methane from oil and natural gas drillers, chemicals used in millions of appliances and tailpipe emissions from trucks, railroads, and even airplanes. (13) Predictably, Republicans in Congress criticized the plan as an abuse of presidential authority, a great threat to the American economy, and pledged their absolute resistance to it.

Despite the temptation to blame the lack of progress toward a significant and necessary climate policy on partisan opposition by Republicans, we must always remember and understand that this is not the only thing that stands in the way of meaningful progress in climate policy. An additional and equally important variable that should never be forgotten is the fact that both Democrats and Republicans, whatever their respective views on climate change, are supportive of policies that will continue to incentivize the very practices that have contributed the most to the problem, oil and natural gas production, and the fact that the policy measures actually being discussed or proposed may be too gradual and ineffective when it comes to actually lowering the accumulation of greenhouse gases. They may only compound the difficulty in the years ahead. Combined with a political dynamic that is inhospitable to any forward-looking policy initiative whatsoever, these obstacles add up to a pessimistic outlook for the possibility of the United States any time soon adopting and implementing a national climate policy. But I would like to suggest a slightly different slant on the discussion that might explain both why the obstacles to progress on the policy front are so persistent and what it may ultimately take to overcome them in the final analysis. It is not at all certain that we will ever overcome all of the obstacles to a sensible policy approach to the climate crisis, because what we will need to do to overcome is well beyond business and politics as usual.

Combating Incremental Failure

Most policy making and decision making in the United States is incremental in nature. It is, we might say, the normal mode of procedure. In practice, incremental decision making considers only a few alternatives for dealing with what is identified as a public problem, and these alternatives will differ only marginally from existing policies. The reasons for this are twofold. First is the matter of efficiency. Because an exhaustive analysis of the costs and benefits of every conceivable option for dealing with a problem in public policy is often unduly time-consuming and expensive, policy makers typically resort to a practical shortcut in deciding on possible improvements or alterations of existing programs and policies. Thus, only a few of the many possible options are seriously examined, and these tend to be ones that involve only small changes in existing policies or procedures rather than radical innovations. Secondly, existing policies and practices are supported by vested and powerful interests that often have a decisive influence on the policy process and the politics that drives it. This frequently means that it can be politically impossible to promote significant shifts in policies or procedures without creating powerful resistance and conflict that can be politically risky or damaging to the policy maker. It is, as a practical matter, easier and less politically risky to build consensus around small adjustments or tweaks in existing policy than to promote new options that deviate from current practices.

From the incremental point of view, and that is the preferred point of view in the American political system, there is no single "right" or "ideal" solution for a problem. Rather, a good or desirable decision is one where support can be built such that a majority of the policy actors can agree on a policy action and adopt it without agreeing that the selected option is the most appropriate or optimum solution. In other words, policies in an incremental process are the product of give and take and the building of a fairly broad consensus among all participants or partisans in the process. It is considered to be politically expedient because it is easier to reach agreement when the matters in dispute are limited to modifications of existing policies and approaches and the policy outcomes are limited, practical, and acceptable to enough policy actors to be adopted. Incremental policy making is fundamentally remedial in nature, in that it is focused on efforts to ameliorate present, concrete imperfections with minor adjustments rather than promoting major future social goals with new policies addressing matters of great value that produce conflict.

Under normal circumstances, assuming a rationally functioning policy process, incremental tinkering is often a satisfactory and very successful approach to policy making. Where the problem to be addressed is not outside the range of the usual matters that require policy adjustments, and assuming that the possibility for deal making exists and incremental adjustments in existing policy can be reasonably tweaked to address a problem, the incremental approach can work out quite nicely. But with respect to global warming, we are not dealing with a normal situation where any of the normal adjustments will work. Likewise, we are not in a political environment at the moment where U.S. policy actors are willing to make deals on even minor incremental adjustments, much less on a problem that is well outside the normal range for standard or typical remedial actions.

Climate change is a post-normal situation, and all of our normal adjustments are failing. As we noted in Chapter 5, in what we might call a post-normal warming climate, many of our conventional assumptions and practices have outlived their usefulness. In a post-normal context, order, we said, seems to be replaced by chaos. People begin to feel that the universe, as they experience it, is no longer rational and orderly. With the impacts of a warming climate already being felt and with the predictions we can make about its future, what has been considered normal and orthodox is not working anymore. What we might consider normal is contradicted by new experiences and changed realities. The experiences we have already had and the knowledge we have gained call into question the adequacy of our business-as-usual approach. We must do more than adjust. The post-normal climate requires significant changes in our thinking and in our normal procedures. It requires, in fact, the creation of a new set of procedures and the establishment of a new normal with respect to how we will live our lives. As we noted in Chapter 5, with respect to climate change we need to stop trying to fit square pegs into round holes. We need to recognize and respond to rapidly changing circumstances. This is to say that, *with respect to climate change and the policies that might address it, the incremental policy approach is an inevitable failure.*

Climate change is exactly the sort of policy challenge that is least suited to an incremental approach. It is simply too big and too complex for business (and politics) as usual. It is a problem that is so massive that it cannot be divided into small portions and be addressed with small adjustments or tweaks in existing policies. The suggestion here is that climate change is not approachable in incremental or gradual steps. What is required is a large-scale and comprehensive policy approach. Incremental federal steps (some increased regulation, cap and trade, carbon tax) and incremental state steps (including adaptation plans) that might be possible are not going to be able to keep pace with the acceleration of risks and vulnerabilities or to significantly slow down the accumulation of negative impacts. Indeed, absent both national and global policy efforts that are much more comprehensive and aggressive to address the causes of global warming (i.e., greenhouse gas emissions), none of our present mitigation or adaptation measures will be of much help.

American policy makers who have spoken on the need to respond to the climate challenge have gravitated toward politically popular and incremental plans such as energy efficiency regulations, renewable fuel standards, setting targets for reducing the rate of greenhouse gas emissions, and some support for alternative energy sources. In addition to spreading things out over what may prove to be too many more decades and back-loading serious efforts to actually reduce the cumulative amounts of carbon and greenhouse in the atmosphere (we will keep adding even as the tub continues to overflow), it is entirely possible that incremental measures can make things worse if they waste time and resources that should have been spent on something far more comprehensive in nature. In other words, time may not be an ally. It might in fact be running out. Additionally, a national policy effort is clearly needed to address the climate crisis, but this need presents itself to us at a time when the U.S. national government has proven to be too weak and divided to act. Those who have created the problem are still being incentivized to make it worse, and too many of our policies aid them in direct contradiction to any of our national, state, or local efforts aimed at mitigation or adaptation. Of course, the failure of the international community to agree on a coordinated and more aggressive course of action handicaps our efforts as well. But thinking strictly of American national policy and the coordination of American efforts, we might ask what is the solution to the almost guaranteed failure of too little too late, too many contradictory policies, and too much back-loading. What is the alternative to ineffective incremental policy tweaking?

What is needed to address a problem as big as climate change is so fundamentally different from our normal mode of thought and action that it seems to be an utter impossibility at first glance. We need an approach that reaches a rational decision that most effectively achieves a given end that will address the causes of the problem. We need an approach that optimizes our outcomes and that leads to the best decisions that will make the necessary large and/or small changes in our public policy. This means the consideration of all alternatives and understanding their costs and benefits in relation to the primary goal to be achieved. This means *separating the problem (climate change) from lesser problems and prioritizing it for action above all else because of its centrality to our national (and international) interests.* Yes, this sounds daunting and perhaps more than a little unrealistic. But this has, though not very often, happened before in the United States.

In 1961 President Kennedy established the goal (national priority) of landing a man on the moon and returning him safely to the earth before the decade was out. At the time this monumental goal was set, the technology to achieve it did not exist, and it was hardly a certainty that it was achievable under even the best of circumstances. Business as normal would have to be changed to meet new challenges. There was more than a little skepticism about the priority and the timeline. But working in a rational, comprehensive manner, the goal was elevated and prioritized. A remarkable national effort was soon under way. Why was this effort undertaken? Mainly because Americans, across partisan lines, felt that this national priority (landing a man on the moon) was made necessary by the crisis known as the Cold War. Beginning with Sputnik in 1957, American fears about Soviet supremacy in space and what it would mean generated a crisis atmosphere and a universally felt need to respond. It was accepted that American technological superiority, including superiority in the space race, was an essential necessity in relation to our most vital national security interests. America's position as a world leader required it to lead the way in space exploration (keeping the moon safe for democracy?) for the benefit of all humankind. Such was, in one form or another, the logic. The cost of doing this was very, very great. But the benefits were much greater in terms of knowledge gained and in terms of the scientific and technological advances made possible by the space program. This created and made possible new solutions to worldly needs and problems, ignited new dreams for the future, and produced incredibly positive economic spinoffs. When the goal was accomplished in July of 1969, a boring incremental "normality" returned. The Soviet Union did not meet the challenge or continue to pose a perceived threat in space. The space program quickly regressed back into the normal or incremental mode in the policy process and, its many subsequent and wonderful accomplishments aside, never again soared quite as high as a policy priority. Indeed, the space program is now seemingly a forgotten stepchild rather than a crowned jewel within the American policy arena.

It has been a long time since 1961. But today the challenge and the objective threats posed by climate change are far more serious than any of the objective threats presented by the Cold War. Now, as then, it may be a moment for setting bold new national (and international) priorities. A state of crisis must be seen to be at hand. The science, the predictable threats and vulnerabilities, the already felt and the anticipated impacts of a warming climate make that case quite convincingly. The crisis is here. As we have said throughout the discussion in this book, the moment of decision is already at hand. With respect to climate change (a.k.a. global warming), we have reached a critical event or point of decision that, if not handled in an appropriate and timely manner, or if not handled at all, will turn into an inevitable disaster or catastrophe of devastating proportions.

Whether it will ever happen or not, I am convinced there can be little forward policy movement of the dramatic sort required until people across partisan lines come to see climate change as a national and global crisis. We must approach the climate race, and we are racing against time, with as much focus and commitment as we did the race to the moon. Sadly, it may take the climate equivalent of a "Sputnik," in this case a monumental climate-related natural disaster, to move the public to perceive that a crisis does indeed exist. Some would in fact suggest that we have already experienced a number of climatological Sputniks that we have ignored. If a profoundly more serious event is required to move public opinion, let us hope it happens before it is, in fact, as I increasingly suspect it will be, too late to address the causes and the effects of a warming climate. Of course, if we begin to understand what science has already shown us and examine the experiences we have already had a bit more closely, we may not find ourselves in need of a "climate Sputnik" to jar us into action. However we come by it, *broad-based public awareness of climate change as a crisis is a prerequisite for meaningful national and international policy initiatives.*

When the American people recognize that a crisis exists, the national government acts. The race to the moon is but one example of several that demonstrate how quickly the United States can summon the unbreakable will and find the resources to create the technologies and the skills required to achieve the "impossible." As a policy matter, the problem of climate change is not a scientific or technical matter; it is a political matter. Our elected officials rarely move with urgency, intelligence, or efficiency on normal issues. Well, climate change is not a normal issue. As the American public comes to see it as a crisis, as they begin to move it to the top of their agenda, their leaders will follow them. Until that time, business (and politics) as usual will provide no incentive for the policy process to move forward.

As we have seen in this chapter, the policy options most under discussion are not new. They have included the carbon tax, the cap and trade, and some moderately increased regulation of emissions. Also mentioned frequently are incentives for alternative, renewable, and cleaner energy sources. Most of the proposals take a gradualist approach, over many decades, and seem to defer serious action until later. Yet none of these options has generated much consistent bipartisan support at the congressional level or resulted in meaningful new legislation. In the meantime, we see that the modest regulatory efforts undertaken at the federal level and the mitigation and adaptation at the state level are not, despite the obvious good they are in fact doing, making a big enough dent. In large part this is because of the many policies that incentivize and support the very things that we have done and are still doing to create the problem in the first place. What policy action might be possible under these circumstances is further diluted by the uncompromising absolutism of the Republicans in Congress toward the whole concept of climate legislation. The Republican Party is the denial party in American politics. Some reject the science altogether. Most Republicans are really probably more opposed to the solutions than they are to the science. They really don't care much about science one way or the other. They don't know very much about it and probably don't feel they need to know much about it. One suspects that some of the attacks on the science are simply a necessary means in their minds to prevent the solutions they abhor on ideological grounds. The reality of climate change and its threats matters less in their political calculations than the short-term political advantage to be targeted, fought for, and won.

If we understand that stabilizing and ultimately reducing CO_2 and other greenhouse gas accumulations in the atmosphere is the *only* possible way to address the causes of climate change, then our mitigation strategy requires more than very gradually decreasing the amount of new emissions by setting targets for the gradual reduction of emissions rates. It requires that we stop relying on fossil fuels for our energy needs. Thinking logically, the *best policy options* that would achieve the optimal result with respect to climate change mitigation are those that *will leave fossil fuels in the ground, promote much more rapid development of alternative and cleaner sources, and establish the bold goal of decarbonizing our energy economy within the next 20 years.* In conjunction with policies aimed at these objectives, things like a carbon tax will be more effective and much more comprehensive regulation will be required. An assumed first step along the path to these best options would be the immediate elimination of all governmental incentives and subsidies for the fossil-fuel industry. The chances of doing any of these things may not be very great at the present time. And this has nothing to do with science or any debates about it. Our politics and our government today are both beholden to corporate interests. Too many of our elected officials have become the willing agents of corporate business plans or, at a minimum, see no difference between a corporate business plan and public policy. Corporate business plans are not about to promote energy and economic policies that address the climate crisis. Corporate and energy interests do not oppose the science as much as they simply want the trillions and trillions of dollars of fossil-fuel profits still in the ground. The threats this poses to people and to the earth's ecosystems are market-externalities they want to simply ignore. That's business as usual. Absent any policy interventions that interrupt the normal flow of their activity, energy interests will pursue profits without giving a second thought to the costly carbon externalities they produce on their way to the bank. At the present time, the prospects for any policy interventions are practically nonexistent.

The impediments to advances in climate-related policies are huge. Any sober assessment of the past 40 years of climate politics and the current policy environment makes it difficult to envision swift progress with respect to climate policy. This does not mean that we cannot prevent the worst impacts of climate change, can't mitigate it, and can't adapt to it. It means only that there is reason to doubt that we will do these things today, or that we will do them in a timely fashion when they will do the most good. We actually know what to do, and we have the ability to do it. The question is whether we will do it. That is a political question. The incremental climate policy illusion is that progress can be made without upsetting the prevailing political and economic interests. The reality is that genuine progress will require that we break free from an incremental approach with its minor tweaks in existing policy and engage in some comprehensive thinking that leads to transitional policies that extend beyond the business- and politics-as-usual mentality. The business-as-usual mentality was great for creating the crisis, but it is proving to be perfectly useless for solving it. But the business-as-usual mentality will persist unless or until the American public comes to see climate change as a crisis. To paraphrase a very wise man, Albert Einstein, all of us will have to use a different kind of thinking than we have used to create the problem.

References

1. Burtraw, D., and Woerman, M. (2012). US Status on Climate Change Mitigation. http://www.rff.org/RFF/Documents/RFF-DP-12-48.pdf (accessed July 14, 2014).

2. Congressional Budget Office (2008). Policy Options for Reducing CO_2 Emissions. http://www.cbo.gov/sites/default/files/cbofiles/ftpdocs/89xx /doc8934/02-12-carbon.pdf (accessed July 15, 2014).

3. Environmental Protection Agency (2009). Acid Rain Program. Environmental Protection Agency. http://www.epa.gov/airmarkt/progsregs /arp/index.html (accessed August 28, 2015).

4. Executive Order—Preparing the United States for the Impacts of Climate Change (2013). http://www.whitehouse.gov/the-press-office/2013 /11/01/executive-order-preparing-united-states-impacts-climate-change (accessed July 14, 2014).

5. Executive Office of the President (2013). The President's Climate Action Plan. http://www.whitehouse.gov/sites/default/files/image/president 27sclimateactionplan.pdf (accessed July 17, 2014).

6. Plumer, B. (2012). "GOP Platform Highlights the Party's Abrupt Shift on Energy, Climate." *Wonkblog. The Washington Post*, Aug. 30, 2012. http:// www.washingtonpost.com/blogs/wonkblog/wp/2012/08/30/gop-platform -highlights-the-partys-drastic-shift-on-energy-climate-issues/ (accessed July 15, 2014).

7. Freeman, J., and Konschnik, K. (2014). "U.S. Climate Change Law and Policy: Possible Paths Forward." Regulatory Policy Program Working Paper RPP-2014-13. Cambridge, MA: Mossavar-Rahmani Center for Business and Government, Harvard Kennedy School, Harvard University.

8. Compston, H., and Bailey, I. (2013). "Climate Policies and Anti-Climate Policies." *Open Journal of Political Science* Vol. 3, No. 4, 146–157.

9. Savitz, E. (2013). Government Subsidies: Silent Killer of Renewable Energy. Forbes, 2/14/2013. http://www.forbes.com/sites/ciocentral/2013/02/14/government-subsidies-silent-killer-of-renewable-energy/ (accessed July 21, 2014).

10. Victor, D. (2009). "The Politics of Global Fuel Subsidies." Global Subsidies Initiative, International Institute for Sustainable Development, November. Available at http://www.iisd.org/gsi/sites/default/files/politics_ffs.pdf (accessed August 28, 2015).

11. Intergovernmental Panel on Climate Change (2014). Fifth Assessment report. http://ipcc.ch/report/ar5/index.shtml (accessed July 23, 2014).

12. Malakoff, D., and Kollipara, P. (2015). "By 98 to 1, U.S. Senate Passes Amendment Saying Climate Change Is Real, Not A Hoax." *Science Insider.* http://news.sciencemag.org/climate/2015/01/98-1-u-s-senate-passes-amendment-saying-climate-change-real-not-hoax (accessed February 20, 2015).

CHAPTER 7

Time Is Not on Our Side

Later Than We Know and Worse Than We Think

It is later than we think and worse than we know. These words when spoken or written in reference to global warming inevitably result in accusations of alarmism. Be that as it may, it is beginning to appear that it may be quite logical to be more than a little alarmed. As we have seen in the discussions in preceding chapters, the fifth assessment report of the Intergovernmental Panel on Climate Change (IPCC) and the most recent U.S. National Climate Assessment both reach this conclusion. (1, 2) Scientific research is telling us, according to these assessments, that the impacts of climate change, those already felt and those to come, are worse than previously understood. They are happening sooner. We are quite possibly closer to irreversible climate change, if it is not already happening, than previously thought. An urgent and rapid global response is a logical requirement. (1)

In addition to things that are happening sooner, the problem itself is actually worse than we might think. In Chapter 2 and our discussion of the science of global warming, we noted that the amount of CO_2 absorbed by the oceans is increasing by about 2 billion tons per year. (3) This absorption, which causes a warming of the ocean waters and slows the surface warming of the planet, is an important contributing factor that will influence the future impacts of atmospheric concentrations of greenhouse gases. Considering that it takes 30 to 40 years for global warming to catch up with atmospheric concentrations of gases, in large part because of the ocean's thermal delaying effect or absorption that slows the process, we have not yet begun to fully appreciate just how bad things really are. Today we are in effect experiencing the impact of atmospheric concentrations of greenhouse gases from the 1970s and 1980s. **In the last 30 years, we have emitted as many greenhouse gases as we emitted in the previous 236 years.** Because of the climate lag, or the ocean's absorption and delaying effect, we have not begun to see the warming that the recent increases in greenhouse gas emissions will ultimately produce. As much as we may be feeling the heat and the impacts of global warming right now, it is going to get much worse because of what we have already emitted into the atmosphere. (1, 3)

Scientific American reported as early as 2012 that the first four IPCC reports tended to underestimate the impacts of climate change. (4) It noted that through all of its work, the IPCC had consistently understated the rate and the intensity of climate change and the dangers of its impacts. Examples abound. Antarctica is losing ice 100 years ahead of schedule. Arctic sea ice is declining 70 years ahead of schedule. Over the past 15 years, sea level rise has exceeded earlier IPCC projections by 80 percent. CO_2 emissions are worse than what the IPCC had previously described as its worst case scenario. Total emissions since 1987 have increased by 81 percent. Finally, the already felt impacts of global warming have exceeded previous IPCC projections. (1, 4) In other words, IPCC projections have proven to be imperfect. This is not, as deniers and contrarians assert, because they overstated the problem or exaggerated the threat. It is because they have consistently understated the problem and erred on the conservative side in making their projections. The reasons for this are quite logical and understandable.

The conservative bias of previous IPCC projections can be attributed to several factors. First, scientists are generally averse to rushing toward dramatic conclusions. Their inclination is to study the matter in much greater detail and demand much more painstaking independent verification before reaching a dramatic conclusion. The more dramatic and potentially controversial the findings, the more painstaking and deliberate the process. Second, if there is drama associated with the social, political, or economic implications of a scientific conclusion, scientists will be all the more deliberate and cautious to assure that their conclusions will be understood and not discounted for nonscientific reasons. Finally, the IPCC process may be even more inherently conservative because its job is to determine where there is scientific consensus, reflect the complete diversity of all valid scientific views where there is not a consensus, and it must in the end produce a summary for policy makers condensing the science even further and incorporating line-by-line revision by representatives from more than 100 world governments—all of whom must approve the final summary document.

Given that the IPCC process is inherently conservative in its estimates, and given the tendency to understate dramatic conclusions that pervades all science really, those who accuse either the IPCC or climate scientists of over-projecting or being alarmist are simply and completely wrong. This also makes the fifth IPCC assessment all the more compelling for its conclusions. The latest IPCC assessment concludes that anthropogenic climate change resulting from CO_2 and other greenhouse gas emissions is irreversible even on a multicentury to millennial scale. The impacts of climate change are being felt everywhere, in every part of the globe, right now. They are more intense than previously projected, and they are accelerating. A rapid global response will be necessary to stave off the worst impacts from global warming. Most importantly, more dramatic action than we have thus far been willing or able to take will be needed to address the causes of global warming. According to the new IPCC assessment, the only way to prevent irreversible damage is through the net removal of greenhouse gases (especially CO_2) from the atmosphere. This means much more than gradually reducing the emissions rates over many decades. It means reducing emissions rates by more than 100 percent of the current annual rate and taking more out than we are putting in every year. Given the inability of governments to achieve major international climate agreements that go beyond modest emissions reductions, the opposition of nearly every energy corporation to really aggressive climate policy of any kind, and the general lack of public knowledge or enthusiasm for dramatic action, one might be inclined to expect a rather catastrophic outcome. That expectation may not be unreasonable in the least.

An Impending Climate Disaster

Climate change (global warming) is a crisis. That is an inescapable conclusion if we understand the science (Chapter 2), its already observed and experienced impacts, and the risks and vulnerabilities it portends for the future (Chapter 4), and if we consider rationally the things that must be done (or are being done, but in too few cases) to address it (Chapter 5). It is a crisis that may very well become a disaster of monumental proportions if we do not respond to it. This disaster would not be the work of nature. If we understand the causes of climate change (Chapter 2), the politics of climate change (Chapter 3), the confusion of public opinion about it (Chapter3), and the repeated failure of policy makers to address it consistently and intelligently (Chapter 6), the crisis could very well lead to a climate disaster that is entirely of human making.

As stated throughout this book and reiterated here, climate change is already a crisis in motion. It is, in fact, moving along or escalating much more quickly than we have previously thought. With respect to climate change, it is clear that the science is telling us that we have arrived at a critical moment or point of decision that, if not handled in an appropriate and timely manner, or if not handled at all, may turn into a disaster or catastrophe. That is now the inescapable conclusion of the fifth IPCC report and the recent U.S. National Climate Assessment. Climate science and its measurements continue to become more precise and more reliable. As the science becomes more precise and reliable, it also becomes perhaps more terrifying.

Throughout our discussion in this book, I have incorporated what I have called an emergency management perspective into the analysis. Emergency management looks at the world differently. It begins with the understanding, for example, that natural disasters are not entirely the work of nature. Nature will do what it will do, of course. This includes events that produce massive storms and other natural phenomena, including floods, droughts, and wildfires. But in most cases, the damages caused by these natural events are human made. In other words, natural disasters are the predictable results of interactions between the earth's physical systems or natural environment; the social, economic, and demographic characteristics of the human communities in that environment; and the specific features of the constructed environment. Because natural disasters and the damages they may cause are predictable, meaning they are not unexpected surprises, emergency managers have learned and put into practice mitigation strategies and adaptive capacities that may reduce the potential for damage impacts from natural events or extremes. In this sense, emergency management has come to mean taking responsibility for disasters. In similar fashion, the conclusion I have reached at the end of this discussion and analysis is that it is clearly time for us to take responsibility for climate change.

Taking responsibility for disasters generally means understanding the three major influences that determine the level of risk and potential cost of any disaster damages. These influences (the earth's physical or natural systems, the demographic characteristics of the human community, and the built or constructed environment) are both often the roots of the problem and the *places where the solutions must be sought for managing the problem.* In almost every case, losses to disasters may be prevented or reduced with forward-looking thinking and planning that accommodates nature, builds resilient communities, and designs a sustainable fit among nature, human beings, and the constructed environment. Taking responsibility means designing that fit and mitigating, adapting, and being smart enough to be forward thinking and proactive instead of merely reactive.

The physical environment is, of course, constantly changing. Being aware of its changes and designing our communities so that they consider those changes and act in harmony with them to whatever extent possible is a practical and necessary thing to do from both a rational and an emergency management perspective. The warming of the climate is a necessary and critical variable to consider in this context. As I have noted from the beginning, that is the context that most interests me and that has inspired much of the discussion throughout this book. Global warming is already resulting in more dramatic meteorological events. In each region of the United States (Chapter 4) and around the globe, this is the case. Storms, floods, extreme temperatures, drought, wildfires, and all other natural hazards are being altered. These alterations mean that the courses of even those disasters that can be expected to recur on a repeating cycle are subject to dramatic changes with respect to frequency and, more importantly, intensity. This being the case, it is critical to realize that the past is no longer a reliable source of information for anticipating the future, because the impact of climate change will change the vulnerability profiles of communities as it impacts the projected course of natural hazards and any resulting disaster events. So it is that we must be forward looking in anticipating these changes, and the communities we live in must be broadly engaged in efforts to mitigate future damages through the adoption of policies and strategies to address their causes and/or the implementation of plans to adapt to what can be expected and address their effects. That's the unassailable logic, as far as I see it.

Where human beings or communities do not take the necessary steps to manage for the causes (mitigation) and effects (adaptation) of identifiable hazards, where they do nothing at all or where they engage in practices that exacerbate the causes and expand the effects, we can say that the resulting disaster is not natural at all. It is entirely human made. In such cases, disasters are created by human design. It is in just this sense that climate change or global warming is well on the way from being a crisis, a dangerous threat, to becoming a full-blown human-made disaster. In other words, from an emergency management perspective, it may be logical to conclude that we are, through inaction, postponed action, inadequate action, harmful action, and denial of the problem, designing a future climate disaster. The science is clear, the facts are grim, the future is potentially tragic, and we continue, for the most part and despite any sincere but half-hearted efforts to the contrary, to design a disaster.

A new study pinpoints the probable dates for when cities and ecosystems around the world will experience warmer environments the likes of which will produce a dangerous and "permanent hot" we have never experienced before in human history. (5) In about a decade, Kingston, Jamaica, will be off-the-charts hot. Projections are that other places will quickly follow (Singapore in 2028, Mexico City in 2031, Cairo in 2036, Phoenix and Honolulu in 2043). By 2047, the entire world will be permanently hotter by far than the hottest years experienced over the last 150 years. By 2047, this report suggests the impacts of global warming could well reach a critical and dangerous level that may well exceed our ability to cope. (5) The 2047 date is based on projections of continually increasing emissions of greenhouse gases from the burning of coal, oil, and natural gases. Should the world manage to meet most of its stated targets for gradually reducing its emissions of CO_2 and other greenhouse gases, the study says the worst or most severe warming can be postponed until 2069. (5) But at this time, the researchers in this study have concluded that we are rushing toward the 2047 date. The scientists who conducted this study, along with every other reputable and legitimate climate scientist, are suggesting that *now* is the time to act. The moment of decision has arrived. The crisis is here. Sadly, the path we are on in the United States and globally is leading us mostly to a disaster entirely of our own making.

Climate change, the current warming of the planet, is an entirely human-caused (anthropogenic) phenomenon. It is disputed by deniers and contrarians, but the science is firm, and its conclusions are indisputable. The warming of the climate that we are experiencing, that we will continue to experience with more devastating effects and that is at the heart of the most threatening hazards and the most incredible vulnerabilities yet known to humanity, is the product of our reliance on fossil fuels for meeting our energy needs. As we adapt our communities to the threats of a rising sea level or the risks of wildfires or the disruptions of our agriculture or the inevitability of cascading failures of our infrastructure, we are in effect adapting them to the hazards and threats we have created through our reliance on fossil fuels. We have designed these threats and produced these hazards. Logic would dictate that, in addition to adapting to these threats of our own making, we might actually want to *stop designing a climate disaster.* Ironically, at a time when the most sensible thing to do might be to set the goal of decarbonizing our energy economy over the next 20 years, the United States is providing more and more incentives for the production of oil and natural gas as one of its top policy priorities. But the logic that tells us to take responsibility for climate change is weakened by our politics. The necessity of addressing a climate crisis is lost against the power and momentum of the fossil-fuel industry and our elected representatives, who are determined to serve their interests above ours. That remains, and is likely to continue to be, the major obstacle to any progress on the climate policy front.

After decades of debate, denial, and delay, more people are coming to understand that we are addicted to fossil fuels that create large quantities of carbon dioxide and that this has been the primary cause of global warming. This addiction also extends to a wide variety of industrial and agricultural practices that create a dangerous amount of other greenhouse global warming gases. We, not nature, have fundamentally changed the climate of the planet we live on. The evidence is mounting, and despite all efforts to deny it or ignore it, we are feeling the heat. The science has warned us for decades that global warming is real and its impacts will be increasingly negative and costly. We have all known that things like energy conservation, renewable energy, reduction of CO_2 emissions, and other measures to mitigate the causes of global warming should be at the top of our national and international policy agenda. For over four decades, we have chosen to wait and see, and the cost of not acting earlier is mounting very rapidly. In recent years, as we have been forced to address the effects of a warming climate, we have all become more aware of the need to invent and invest in costly adaptive measures to protect our communities and their infrastructures form the effects of a warming climate. Even with this growing awareness, we have only just begun to adapt. The question is whether we will actually be able to do enough adaptation to stay ahead of the problem. Our failure to take mitigation seriously, at least until now, has allowed the problem to become much worse and has contributed to higher adaptation costs and created more daunting challenges and threats for the future. Will we at long last be ready to act with a greater sense of urgency?

Despite any elevation in the level of public concern about the climate crisis—and there still remains great division and misunderstanding in the body politic—the ongoing soap opera in the U.S. Congress is scripted to prevent much in the way of progress on the policy front. In 2015, in the first 100 days of the 114th Congress, the top priority of the Republican majority became abundantly clear in a series of roll call votes to undermine environmental protections or fast-track projects like the Keystone XL pipeline. Most of these efforts did not result in new law. For example, legislation to approve the pipeline, rather than wait for the State Department to issue its final national interest report on the project, was passed by Congress in February but ultimately vetoed by the president. (6) One still suspects, or has good reason to fear, that the pipeline will move inexorably forward and be approved in the end. Beyond that, it is clear that the primary goal of the Republican majority in Congress is to undermine environmental laws and prevent any meaningful action whatsoever to address the climate crisis.

As the United States and nations around the world prepared for the Climate summit in Paris in December of 2015, amidst some signs of renewed hope for an international agreement on reducing carbon emissions, President Obama announced the U.S. plan to reduce CO_2 by 26 to 28 percent over the next decade. This would be done through existing regulatory authority. In essence, this is a continuation of the Obama strategy to do what can be done through executive action. There are no new policy initiatives on the horizon and absolutely no reason to believe there will be any time soon. Kentucky Senator and Republican majority leader Mitch McConnell warned the UN and the international community to "proceed with caution" in negotiating a climate deal with the United States because the country (if he and the Republican Party have their way) would not make good on Obama's commitment to reduce emissions. (7) McConnell's statement came just hours after the U.S. government submitted its pledge to the United Nations. Such prompt and total rejection of the U.S. proposal, and a not-so-thinly-veiled threat to thwart the effort to implement it, are situation normal in the U.S. Congress where, unfortunately, the climate policy train is off the rails. This is what designing a climate disaster looks like in the U.S. Congress. This is the oppressive political reality that significantly dampens any prospects for addressing the climate crisis in a timely and intelligent fashion.

Concluding Remarks

It would be presumptuous of me to suggest that any of the conclusions that I have reached in my efforts to study and understand climate change are unique. It would be equally presumptuous to suggest that the path forward is easy or that I have any certainty about the prospects for the timely adoption and implementation of concrete solutions to address the problem, the crisis known as climate change. What I hope to have accomplished with this book is in the end very basic. The effort herein has been directed at explaining the scientific evidence that supports the definition of climate change as a serious problem, as a crisis already upon us. It has further attempted to explain the politics over the past 40 years that has distracted American citizens from an understanding of the problem. As a part of that effort, an examination of American public opinion and the things that have influenced it was considered essential. This is premised on the political reality *that for any significant policy progress to be made to address the climate crisis, broad public support will be necessary.* A careful articulation of the risks and vulnerabilities that a warming climate imposes on all communities in all regions of the United States was provided in the hope that it might make the global climate challenge more personal and local, thereby enhancing our understanding of it as a part of our individual experience and as a direct influence on the quality of our lives. Next, a discussion of mitigation and adaptation strategies was engaged to demonstrate the sorts of things that can be done and are being done, but in too few cases, to address the problem of climate change. Finally, an examination of the current American policy environment and the prospects for meaningful action demonstrated that we are far behind where we urgently need to be in addressing the problem.

If I were to write a single paragraph describing the discussion we have had in this book and the conclusion to be reached in the end, it would be the following. We have seen that the climate change or global warming debate is not a scientific debate. The science is compelling and conclusive, and the consensus it has produced is undeniable. Our politics in the United States is too often—and by corporate and partisan design—totally out of touch with scientific reality. As a result, public opinion is confused, and our thinking is unclear. The measurable impacts already felt and the threats and vulnerabilities presented for our future are mounting. Time—though no one can pinpoint the moment where an irreversible tipping point will ultimately be reached—is not really on our side. Things are moving more quickly, and the costly consequences of inaction are arriving at our doorsteps ever more convincingly. Every moment already wasted, and every moment we continue to waste in responding to the growing climate crisis, is a building block in a mega-disaster of our own making. The American public in particular must be both more involved and better informed on the subject. We ignore it at our great peril, debate it to the point of moral absurdity, and tolerate inaction or obstruction by our elected officials at an astonishing and unsustainable cost to ourselves and the generations to come.

If there is one genuine surprise I have experienced in the writing of this book, it is that I am willing to run the risk of being called an alarmist on the subject of climate change. A sober analysis of what the science has shown me combined with my analysis of how our political process has constructed and continues to live in an alternate reality quite different from the scientific reality has had an effect on me. My emergency management orientation means that I am inclined toward the objective identification of risks and vulnerabilities. A logical and necessary corollary to risk and vulnerability assessment is the taking of logical steps to reduce the causes (mitigation) and/or anticipate and manage the negative effects of a hazard threat (adaptation). Given that general orientation of mind, it is impossible for me to look at the climate crisis and not to see a logical progression that is supported by all of the scientific evidence. Global warming is happening. It is happening because of our reliance on fossil fuels for energy and our emission of greenhouse gases into the atmosphere. This is a bad thing. It is causing and will continue to cause great damage. There are things we can do to address the problem. We can mitigate; we can adapt. To have seen that logic in relation to climate change, and to have come to the conclusion that we in the United States are doing next to nothing about it has caused me to be . . . well . . . alarmed. This is especially the case when we look at what is being projected for the future. It really is later than we know, and it is worse than we think. I have concluded that to prevent the greatest catastrophes associated with a full-fledged climate disaster, we must convince the American public and our leaders that it is time to act decisively, comprehensively, and immediately. Simple emissions reductions drawn out over many decades will not be enough. We must do much more. We must do it now. The completion of this book has led to the intensification of my own sense of urgency about acting.

The partisan values and practical concerns that divide us on climate change must not prevent us from seeing that the future of humanity requires a new kind of thinking generally. Everything we know and that science has made clear about the risks and vulnerabilities associated with a changing, warming climate should at a minimum lead us to agree on five basic principles. These principles are based on the urgent need to make our social, economic, and political efforts more relevant for a world in which a warming climate is dramatically and negatively altering the future for humanity. Managing the crisis means managing the risks. Reflecting on the discussion we have had throughout this book, these are basic principles that must be embraced by all relevant public and private actors. Whatever our individual orientation to the subject matter, the science that it is based on, or our partisan preferences, these five principles stated in general terms should guide our discussions and all of our public and private actions, if we are at all logical. Even if we will continue to disagree about strategy and specific policy options, these principles must orient the discussion and all of the debate.

The *first principle and the first priority* must be to **decarbonize our energy economy**. To do this, we must ideally promote policies that do two necessary things as quickly as feasible. We must encourage the production of renewable energy sources, and we must promote policies that reduce carbon emissions immediately. Addressing the threats associated with climate change requires that we focus on the source—carbon pollution. None of the answers we seek will be found in the continued extraction of fossil fuels or the adding of more carbon pollution by new and presumed more-efficient bridge fuels such as natural gas. None of the answers we seek can be found by geo-engineering or technological miracles that, in addition to posing new and dangerous risks not yet fully understood, promise to remove carbon or block the sun to make continued reliance on and extraction of fossil fuels a part of our long-term future. It is only by helping society to reduce harmful emissions and by developing carbon-free energy sources that we can take the first meaningful steps to mitigate and reduce the most severe future impacts of climate change. This is the ultimate solution to the climate crisis: to build a secure, clean, renewable, carbon-free energy economy.

The second principle is that we must **plan for the future, not the past**. Even if we make progress in clean energy production and emissions reductions, we will continue to have impacts from a warming climate that will require us to adapt. What we will have to adapt to will be very different from what we have previously experienced. Traditional risk modeling, for example, uses past experience to predict

the impact of future weather events. But in the case of rapid and unprecedented climate change, historic weather patterns may not always
be reliable predictors of future conditions. We will need to re-assess
our risk exposure based on new and emerging climate data. Educating
ourselves about growing climate risks is the first critical necessity to be
met to help American communities to plan for the future.

The third principle is that we must **build our communities to last.**
Climate change and increases in extreme weather events threaten
communities and their basic infrastructure such as roads, bridges,
airports, water treatment facilities, and dams. Policy makers, planners, businesses, and citizens must work together to strengthen the
climate resiliency of communities. For example, the creation and en-
forcement of stronger building codes so that property is secured
against high winds, flooding, power losses, and extreme heat are
practical necessities to be met. Adjustments in land use planning and
ensuring that critically exposed markets are more resilient against the
impacts of climate change and are rebuilt to withstand environmental catastrophes will also be necessary. As a general proposition, this
principle suggests that investing in greater resiliency today will help
to ensure that communities are prepared for the future.

The fourth principle is the determination to **design and develop
climate-conscious products and incentives.** Our transition to a low-
carbon economy will of necessity mean scaling up products and services that promote clean and efficient energy use as well as resilient
building design. Innovative products, such as "Pay-As-You-Drive"
auto insurance (PAYD) and discounted premiums for energy-efficient
buildings, provide powerful incentives to reduce energy use and, in
turn, climate changing greenhouse gases. Homeowners and businesses should be educated about weather-resilient building materials
and techniques. They should also be given incentives for implementation of these techniques. Public policies and private practices that
promote the developing of new products and underwriting emerging
clean energy developments such as wind farms and solar power ar-
rays will not only speed up the transition to a low-carbon economy
but lead to economic opportunity and profit as well.

The fifth and final principle is that major economic actors must **invest to manage climate risks.** Investors can choose to invest in clean
energy innovation or continue to bet on climate-warming fossil fuels.
They can invest in enterprises that are leading the way in water conservation, energy efficiency, and environmental responsibility, or they can
support those pursuing "business as usual" in an age of ever-scarcer

resources. Investors need to consider the sustainability of their enterprises and portfolios. In that regard, climate change is no longer an extraneous, nonmaterial financial concern for investors or for the enterprises they invest in. It's a significant risk that must be managed responsibly.

The five basic principles just enunciated presuppose that we understand that a climate crisis is already at hand. It suggests even that politicians and citizens alike have begun to understand that our public policies and our economic priorities will have to change. This means an agreement that the need to mitigate and adapt to the impacts of climate change is viewed by all as an indisputable necessity. That's presupposing and suggesting a lot. But this is ultimately where we need to get in our public discourse before we can begin to implement the five principles. Only then can we begin to manage the climate crisis. How close are we to being ready to manage the crisis?

Optimism is perhaps the preferred note to strike at the end of a book such as this one. But it is difficult to be optimistic in our present policy environment and in the context of the past 40 years of increasingly pointless political skirmishing about climate change. Every day is new and, of course, an opportunity to create a better future. One can say that there are signs that more and more Americans are beginning to understand and that, as a result, the political dynamic may be less of an obstruction moving forward. One has to admit this is possible and work to improve our odds of success in an imperfect world. Yet, with respect to opportunities missed and paths not taken, a day later eventually does become a day too late. It may just be that our inability to act on the climate crisis, despite all that is known, and despite the fact that we do know what can be done and how to do it, will be our greatest failure. A more optimistic projection will not be possible until every citizen in every nation understands and responds intelligently and with an appropriate sense of urgency to what science is telling us. What it is telling us is rooted in evidence that is beyond dispute. The message is clear, and our responsibility to respond is equally clear. We must understand and heed the five most important words of our age: EARTH, WE HAVE A PROBLEM! Then we must do the most important and difficult thing we have ever had to do. We must manage the climate crisis.

References

1. International Panel on Climate Control (2014). Climate Change 2014: Impacts, Adaptation, and Vulnerability: Summary for Policymakers. http://

ipcc-wg2.gov/AR5/images/uploads/IPCC_WG2AR5_SPM_Approved.pdf (accessed July 31, 2014).

2. National Climate Assessment (2014). http://nca2014.globalchange .gov/ (accessed July 31, 2014).

3. Sabine, C.L., et al. (2004). "The Oceanic Sink for Anthropogenic CO_2." *Science* vol. 305, 367–371.

4. Scherer, G. (2012). "Climate Science Predictions Prove Too Conservative." *Scientific American* December 6. http://www.scientificamerican.com/article /climate-science-predictions-prove-too-conservative/ (accessed July 29, 2014).

5. Mora, C., Frazier, A.G., Longman, R.J., Dacks, R.S., Walton, M.M., Tong, E.J., Sanchez, J.J., Kaiser, L.R., Stender, Y.O., Anderson, J.M., Ambrosino, C.M., Fernandez-Silva, I., Giuseffi, L.M., and Giambelluca, T.W. (2013). "The projected timing of climate departure from recent variability," *Nature* 502, 183–187.

6. Valentine, K. (2015). "Congress Has Made Undermining Energy and Environmental Laws the Focus of Its First 100 Days." *Climate Progress.* http://thinkprogress.org/climate/2015/04/15/3647310/congress-first-100 -days-energy-report/ (accessed April 21, 2015).

7. Rogers, K. (2015). "Mcconnell Tells Other Nations Obama Can't Fulfill Emissions Reduction Pledge." *Energy Guardian.* http://energyguardian .net/mcconnell-tells-other-nations-obama-cant-fulfill-emissions-reduction -pledge (accessed April 21, 2015).

Index

About the Author

ROBERT O. SCHNEIDER, PhD, is a professor of public administration at the University of North Carolina-Pembroke. His areas of expertise include emergency management practice and policy. The author of the book *Emergency Management and Sustainability: Defining a Profession,* Schneider has published numerous peer-reviewed articles in the field of emergency management. His interest in climate change is the product and an extension of his work in the study of natural hazard mitigation.